Small-Scale Agriculture in America

Small-Scale Agriculture in America

Race, Economics, and the Future

Ejigou Demissie

Westview Press
BOULDER, SAN FRANCISCO, & OXFORD

Westview Special Studies in Agriculture Science and Policy

This Westview softcover edition is printed on acid-free paper and bound in library-quality, coated covers that carry the highest rating of the National Association of State Textbook Administrators, in consultation with the Association of American Publishers and the Book Manufacturers' Institute.

Published in 1990 in the United States of America by Westview Press, Inc., 5500 Central Avenue, Boulder, Colorado 80301, and in the United Kingdom by Westview Press, Inc., 36 Lonsdale Road, Summertown, Oxford OX2 7EW

Library of Congress Cataloging-in-Publication Data
Demissie, Ejigou.
Small-scale agriculture in America: race, economics, and the future / Ejigou Demissie.
p. cm.—(Westview special studies in agriculture science and policy)
Includes index.
ISBN 0-8133-7823-0
1. Farms, Small—United States. 2. Farms, Small—Government policy—United States. 3. Agriculture—Economic aspects—United States. 4. Agriculture and state—United States. 5. Farmers—United States—Economic conditions. 6. Afro-American farmers—United States—Economic conditions. I. Title. II. Series.
HD1476.U5D46 1990
338.1'6—dc20 89-9117
CIP

Printed and bound in the United States of America

The paper used in this publication meets the requirements of the American National Standard for Permanence of Paper for Printed Library Materials Z39.48-1984.

10 9 8 7 6 5 4 3 2 1

Dedicated to:
Mr. and Mrs. T. J. Henry,
Rev. and Mrs. D. G. Jones,
and my family

Contents

Tables

Preface

Over the past several decades, changes have occurred in the structure of American agriculture that have fostered such a decline in the number of farming units as to practically result in the disappearance of middle-sized family farms. Consequently, U.S. agriculture has become a bimodal agricultural system, with a large number of small-scale farms coexisting with a smaller number but an increasing proportion of large-scale farms.

The structural changes did not stop when farms classified as middle-sized ceased to exist. Rather, those changes began to be reflected in a decrease in the number of small-scale farms as well, particularly those owned and operated by minorities. Unless some action is taken to turn this situation around, the virtual extinction of small-scale farms will not be far off.

Small-scale farmers have traditionally been the backbone of American agriculture and of the U.S. economic and political system. Some historians argue that the very form and philosophy of American government would not have evolved as it has today had the U.S. not been a land of "small" farmers. Today, small-scale farmers are not significant producers of food or fibre. But they are still an important part of the total rural infrastructure, particularly in areas where there are fewer or declining farm numbers. They contribute directly to jobs; to wages and income; to the existence of retail and service establishments, including the financial institutions; and to the rural tax base of the nation.

Apart from the immediate effect on the individual farm families, the decline and economic problems of small-scale farmers raises the specter of substantial long-run changes in the structure of agriculture and community life, particularly in rural areas. These changes include the severe economic problems already gripping many rural businesses and public services, the deterioration of basic physical infrastructure, and rural flight.

What should be done to bring about public awareness of the changes that are taking place in small-scale farming and rural communities? What should be done to win meaningful support for small-scale family farms and rural communities and to move the nation toward a more competitive agricultural economy? My thesis is that the local, state, and federal governments must broaden their approaches to alleviating the problems of small-scale agriculture.

Preserving communities in rural areas is in the best interest of the millions of people who live there. It is also in the best interest of the entire population of the United States, who depend upon rural areas as sources of food, fibre, and forest products. Rural communities are also stewards of the nation's land and water resources. It may be possible to create a rural renaissance if we think about the people, their work, their education, and their sense of culture.

This study analyzes the relationship between small-scale agriculture and the process of development -- or the lack thereof -- in rural areas. Certainly, the scope of this study hardly exhausts the list of pertinent issues, nor does it discuss the issues in the fullest possible detail. What it does propose to do is stimulate dialogue on the issues, with the hope that through discussion, action may eventually be made possible.

This book was written for several audiences in the agricultural and non-agricultural communities, including educators, extension workers, farmers, researchers, and students of agriculture, as well as any citizen who shares an interest in and has a concern for small-scale agriculture and U.S. rural development. I sincerely hope that this book will stimulate its readers to learn more about the role small-scale agriculture plays in our lives. I also hope it will make readers more sensitive to the perceptions, problems, and goals of minority small-scale producers, particularly blacks.

Ejigou Demissie

Acknowledgments

The successful completion of this book required the assistance of many individuals, including the anonymous reviewers, and I thank them all for their help. I specially thank Mortimer Neufville, Dean, School of Agricultural Sciences, University of Maryland Eastern Shore for his total support and for his review of the manuscript. I am grateful to Joe McFarlane for her excellent editorial assistance. I am also very deeply grateful to Kellie Masterson, Lynn Arts, and Mary Beth Nierengarten of Westview Press for their encouragement to pursue the writing effort, their review of the manuscript, and their editorial assistance.

Special recognition is given to Clarice Corbet for the considerable amount of time she devoted to typing the initial draft of the manuscript, and to James Hayes and Laura Sterling for their help in processing the final copy of the manuscript. Last but not least, a very special debt of gratitude is expressed to my wife, Nohora C. Rivero, and son, Dawit E. Demissie, for their support and tolerance, which made the preparation of this book possible.

E. D.

1

Introduction

In the past, most policy efforts to stimulate production in the U.S. were directed toward the large-scale commercial farms.[1] These farms, with efficient use of economic principles, adaption of new technology and substitution of capital for labor, have dominated the production and market shares over the years. As a result, for the past several decades, we have seen changes in the structure of farming characterized by trends toward fewer but larger farming operations.

The census of agriculture data on U.S. farm numbers in Table 1.1 verify these trends in this century. As the table indicates, the decline has been very substantial. For example, the number of farms declined from 5.7 million in 1900 to 3.1; 2.2; and 2.0 million in 1964; 1982; and 1987, respectively.

The decline in number of farms is generally true for both white and black farm operators. However, the decline has been more significant for black farmers than whites. For example, in 1900, of the total 5,737,000 farms in the U.S., blacks represented 746,717 (13%). Today, there are fewer than 23,000 farms operated by blacks in the country, representing about one percent of all farms in the U.S. (Table 1.1).

Land in farms has also shown a decrease over the years. However, compared to the number of farms, the decrease has been gradual. For example, in 1940 there were 1,061 million acres of farmland, compared to 1,015 million acres in 1978; 987 million acres in 1982; and 964 million acres in 1987. Since the decrease in farm numbers was relatively greater than the decrease of land in farms, the average size of farms has

TABLE 1.1: Number of Farms and Acres of Land in Farms, U.S., 1900-1987

Year	Number of Farms (000)	Total Land in Farms (000 acres)	Average Size (acres)	Number of Black Owned Farms
1900	5,737	838,592	146	746,717
1910	6,362	878,798	138	893,377[a]
1920	6,448	955,884	148	925,710
1930	6,289	986,771	157	882,852
1940	6,097	1,060,852	174	681,790
1950	5,382	1,161,420	216	559,980
1954	4,782	1,158,192	242	467,656[b]
1959	3,711	1,123,508	303	272,541
1964	3,158	1,110,187	351	184,004
1969	2,730	1,062,893	389	87,393
1974	2,314	1,017,030	440	45,594
1978	2,258	1,014,777	449	37,351
1982	2,240	986,797	440	33,250
1987	2,088	964,471	462	22,954

[a] Alaska not included.
[b] Alaska and Hawaii not included.

Source: Various issues of the *Census of Agriculture for U.S.* United States Department of Commerce, Bureau of the Census, Washington, D.C.

increased significantly since 1900. The average size of farms for the country as a whole has increased from 146 acres in 1900, to 216 acres in 1950, to 389 acres in 1969, and to 462 acres in 1987 (Table 1.1).

Currently, the U.S. is losing some 30,000 farms annually. If present trends continue, the number of farms is projected to decline to 1.8 million in the year 2000 from what it was in 1982 (1). In fact, a recent study by the Office of Technology Assessment projected the number of farms to decline from 2.2 million in 1982 to 1.2 million in the year 2000 (2). Given that the amount of land in cultivation has not drastically changed, this means that farms will become even larger in size, and ownership will become more and more concentrated in fewer and fewer hands. As a result, 75 percent of the total agricultural production in the year 2000 is expected to come from approximately 50,000 farms (2).

Statement of the Problem

The rate of decline in the number of farms has not been uniform through the agricultural sector of the nation. Most of the decline in the number of farms has been in the small-scale farms[2] category. Particularly the number of small-scale farms declined significantly over the past several decades while the number of large-scale farms continued to increase. For example, in 1987, the number of small-scale farms was 1,511,000 less than what it was in 1964, a decrease of 50 percent in just 23 years. For the same time period, however, the number of large-scale farms increased by 308 percent (Table 1.2). Although the number of small-scale farms has been declining rapidly, today more than 70 percent of all U.S. farms are still considered to be small-scale farms. In 1987, about 72 percent of the nation's farms had sales less than $40,000.

Various combinations of factors have contributed to the structural changes in agriculture, hence to the problems of small-scale farmers. The major factors are changes in technology, capital requirements, government policies and

TABLE 1.2: Distribution of U.S. Farms by Sales Class, 1964-1987

	1964		1974		1982		1987	
Sales Class	Number (000)	%	Number (000)	%	Number (000)	%	Number (000)	%
Less than $10,000	2,288	72.2	1,203	52.0	1,096	48.9	1,028	49.2
$10,000 - $19,999	467	14.8	310	13.4	259	11.6	251	12.0
$20,000 - $29,999	180	5.7	188	8.1	145	6.5	--	--
$20,000-$24,999	--	--	--	--	--	--	76	3.6
$30,000 - $39,999	80	2.5	134	5.8	103	4.6	--	--
$25,000-$39,999	--	--	--	--	--	--	150	7.2
Sub Total	3,015	95.4	1,875	79.3	1,603	71.6	1,505	72.0
$40,000 and over	143	4.5	479	20.7	637	28.4	283	28.0
Total	3,158	100	2,314	100	2,241	100	2,088	100

Source: *Census of Agriculture for U.S.*, 1964, 1982, and 1987, United States Department of Commerce, Bureau of the Census, Washington, D.C.

programs, availability of financial credits, management practices, national market structure, forms of ownership, price instability, and orientation of national research and extension agencies towards small-scale farmers. While it is recognized that these factors have an immediate and dramatic importance for the farm sector, their effects on the structure of agriculture are long-term (1).

This significant concentration in land ownership and production is a serious disadvantage to the rural economy of the U.S. It has given rise to a great deal of uncertainty about the future of small-scale farms as viable economic units and as a way of life for many rural residents. Consequently, the preservation of small-scale farms has emerged as a critical issue in agricultural policy deliberation at the federal, state, and local levels of government (3, 4, 5).

State and federal agencies and other organizations have convened numerous workshops and conferences to discuss problems facing small-scale farmers and to propose possible solutions to those problems. Some programs to improve the small-scale farmers' lot are already underway. But the problems of small-scale farms continue, and many of the small-scale farmers still face the same mix of problems that have driven so many of their fellow farmers out of business. This is particularly true in the Southern region of the country, where the greatest concentration of small-scale farmers in general, and blacks in particular, is found (Table 1.3).

Small-Scale Farmers

Most small-scale farmers are full owners of their land. They have organized as sole proprietors, and they represent about 10 percent of the total agricultural product sales of all farms in the country in 1987. Most of these farmers tend to specialize in livestock (beef cattle) and cash grain enterprises. Furthermore, these farmers tend to be older, live on the farm they operate, have low income, attain a low level of education, work over 100 days a year off the farm and cultivate a small portion of their land. Although the age distribution of these

TABLE 1.3: Number and Percent of Small-Scale Farms in the South, 1987

State	Total Number of Farms	Number of Small-Scale Farms	Percent
Alabama	43,318	36,175	83.5
Arkansas	48,242	34,943	72.4
Delaware	2,966	1,528	51.5
Florida	36,556	28,609	78.3
Georgia	43,552	32,835	75.4
Kentucky	92,453	82,291	89.0
Louisiana	27,359	21,005	76.8
Maryland	14,776	10,787	73.0
Mississippi	34,074	27,479	80.6
North Carolina	59,284	45,440	76.6
Oklahoma	70,228	58,921	83.9
South Carolina	20,517	17,269	84.2
Tennessee	79,711	72,432	90.9
Texas	188,788	157,993	83.7
Virginia	44,799	38,259	85.4
West Virginia	17,237	16,198	94.0
Total	823,851	682,164	82.8
U.S.	2,087,759	1,504,458	72.1

Source: *Census of Agriculture for U.S.*, 1987, United States Department of Commerce, Bureau of the Census, Washington, D.C.

farmers varies by states, in general, fewer small-scale farmers are young (less than 35 years old) and more are older (at least 54 years old). A majority of these farmers (nearly one-half of small-scale farmers in the U.S.) are located in the South (Table 1.3).

Like all small-scale farmers in the nation, black operators are full owners of their farms, and a majority of these farmers tend to be older, have lower annual agricultural sales relative to all farmers in the country, and live (95%) in the South. Among all black farmers in the South, for example, 60 percent are over 54 years old while only 6 percent are under 35 years.

Most blacks engaged in farming in the U.S. operate on small acreage. In 1987, approximately half of these farmers had 49 acres or less and a third had 50 to 139 acres. For example, farm size of black operators in the South averaged 99 acres, compared with 328 acres for all farms in the region.

When size is measured by volume of sales, less than 5 percent of all black farmers in this region sold more than $40,000 worth of farm products, while about half (46%) of these farmers had sales under $2,500. Most of the black small-scale farmers in the U.S. in general, and the South in particular, tend to specialize in tobacco, cash grain, and beef cow enterprises. About 54 percent of these farm operators had a primary occupation other than farming and 36 percent spent 200 days or more working off the farm.

Objectives

Based on data from the Census of Agriculture and the United States Department of Agriculture (USDA), this study's main objective is to describe the history and current status of small-scale farm operators in the U.S. and to provide background information for small-scale farm policy makers and program developers in the nation. The will: (1) describe the characteristics and trends of small-scale farms and farm operators; (2) to examine the history, status, and role of black farm operators in production agriculture; (3) to evaluate the financial trends in the U.S. small-scale agriculture; (4) to identify and discuss the various constraints or factors for increasing farm income on small-scale agriculture; (5) to examine the role of small-scale agriculture in rural development and stability and consider the implications of the constraints on the rural areas; and (6) to examine alternative farm programs and policies as solutions to the problems of small-scale agriculture in the country.

Organization of the Remainder of the Book

The remaining chapters of the book are organized as follows. Chapter 2 shows the profile of small-scale farms with respect to their agricultural production, resource use and control, farm asset holdings, types of enterprises, and farm production expenses. Specifically, the chapter compares and contrasts small-scale farmers with large-scale farmers to show the deterioration in the past and the current status of the nation's small-scale agricultural economy.

Chapter 3 describes the difference between black and white farm operators while examining the history, status, role, and current situation of black farm operators in the U.S. The chapter also examines some factors affecting the existence of black farmers in production agriculture, discussing the history and importance of historically black Land-Grant institutions for the survival and development of blacks in general and for black farm operators in particular.

A description of the financial conditions of the farm and farm operators in the U.S. agriculture is given in Chapter 4 to provide a realistic assessment of the current and future financial conditions of farm operators in the country in general, and of small-scale farmers in particular. The discussion presents a comparison of small-scale farmers by economic class, offering insights into the type of small-scale farmers most affected by the current financial situation.

Chapter 5 reviews some of the constraints responsible for the demise of small-scale farm operations in the U.S. This chapter describes the various combinations of factors that have given rise to the current constraints. Chapter 6 examines the role of small-scale agriculture in the rural development of the country, examining the implication for the development of rural areas if small-scale farm operations are allowed to fade away.

Finally, Chapter 7 examines some programs and policies as alternative solutions to the problems of small-scale agriculture in general and of black farm operators in particular. The chapter suggests some research and extension priorities that would allow local, state, and federal government programs and policies to stop the further decline in the number of small-scale farmers and the deterioration in the agricultural economy of rural areas.

Notes

1. Various definitions of a farm have been used in earlier censuses; however, since 1974, a farm is defined as a farm if it has agricultural product sales of $1,000 or more.

2. In this study, a small-scale farm is defined as a farm with gross sales of farm products of less than $40,000 during a year.

References

1. Lin, W., et al. *U.S. Farm Numbers, Sizes and Related Structural Dimension, Projections to Year 2000.* ESCS/USDA Technical Bulletin, No. 1625, Washington, D.C., 1980.
2. U.S. Congress, Office of Technology Assessment. *Technology, Public Policy and the Changing Structure of American Agriculture.* OTA-F-285, Washington, D.C., 1986.
3. *Regional Small Farms Conference: National Summary.* Co-sponsored by USDA Community Services Administration and ACTION, 1978.
4. Carlin, T. A., et al. "Small Farm Policy: What Role for the Government," in *Increasing Understanding of Public Problems and Policies.* Farm Foundation, Oak Brook, Illinois, 1981.
5. Humphries, F. S. "U.S. Small Farm Policy Scenarios for the Eighties," *American Journal of Agricultural Economics,* 62 (1980).

2

General Characteristics and Trends for Small-Scale Farms and Farm Operators

U.S. agriculture is an industry that has been, and still is, the bedrock of the country's economy, accounting for over 20 percent of the jobs in the nation. The industry is also a major contributor to the country's economy through export sales. These export sales bring dollars into the U.S. and trigger chain reactions by stimulating additional economic activity in many parts of the nation's economy. Overall it accounts for over 20 percent of the gross national product of the total U.S. economy.

The majority of the agricultural revenue comes from the sale of products by large farm operations which represents only 28 percent of all farms in the country. Small-scale farms account for only 10 percent of all agricultural product sales in the country.

General Characteristics and Trends for Farms

Number of Farms and Amount of Land in Farms

The actual number of small-scale farms in the U.S. decreased by about 20 percent, from 1,890,000 in 1978 to 1,504,000 in 1987. As a result, small-scale farms in 1987 represented 72 percent of all farms in the country (Table 2.1). Total land in farms has also been declining throughout the

TABLE 2.1: Selected Characteristics of Small-Scale Farms, 1978-1987

Characteristics	1978	1982	1987
Number of farms (000)	1,890	1,604	1,504
Land in farms (000, acres)	356,911	287,130	297,732
Average size of farm (acres)	188.8	179.0	197.9
Harvested cropland (000, number)	1,484	1,214	1,108
Acres harvested (000)	98,508	71,193	63,924
Cropland per farm (acres)	66.4	58.7	57.7
Cropland used only for pasture (number)	813,576	646,080	612,235
Acres pastured (000)	49,368	39,695	39,044
Woodland (number)	809,041	689,698	611,917
Acres woodland (000)	54,926	47,518	43,608

Source: *Census of Agriculture for U.S.* 1974, 1978 and 1987, United States Department of Commerce, Bureau of the Census, Washington, D.C.

1970s and early 1980s. For example, in 1982, there were 287 million acres of land in farms compared to about 357 million in 1978, representing a decrease of 20 percent for the period. The percentage of the nation's total land area devoted to farming by the small-scale producers increased by about four percent in 1987 over what it was in 1982. However, the number in 1987 was 17 percent less than that of 1978.

The average size of small-scale farms also declined throughout the 1970s and early 1980s. The average acre size per farm in 1978 was about 189, and by 1982 the average size dropped to 179 acres per farm. But, by 1987, it increased to about 198 acres per farm due to a combination of a decrease in the number of farms and a slight increase in land in farms.

Farm Land Use

In 1987, small-scale farmers had relatively more of the land they controlled in harvested cropland. Farms with harvested cropland decreased their harvested acres in 1987 over what it was in 1982 despite an increase in the acres of total land in farms for the same time period (Table 2.1). In 1987, there were 25 percent fewer farms of harvested crop land compared to 1978 and 9 percent compared to 1982. For the same time period, total harvested acres decreased by 35 and 10 percent, resulting in a decrease of 8 and 1 percent of harvested acres per farm, respectively.

The number of farms with cropland used only for pasture (i.e., the number of livestock farms) increased in the late 1970s, but has decreased since then. Obviously, many of the nation's farms have both harvested and pastured crop land, but the decrease observed relates to the number of livestock farms only. This is because the number and pastured land acres of these farms have dropped for the study period. For example, the number of these farms declined by 25 percent while their pastured land acres dropped 21 percent between 1978 and 1987 (Table 2.1).

Farms with woodlands have also declined in number as well as in woodland acres between 1978 and 1987. The decrease in acres in 1987 was about 21 percent over what it was in 1978. Although the number and acres decreased for the study period, the decrease in both the number and acres of woodland farms between 1982 and 1987 was minimal.

Types of Farms and Enterprises

Though not the only crops grown by small-scale farms, the major crops[1] raised by these farms are shown in Table 2.2. As the table indicates, small-scale farms raised eight major crops using about 22 percent of all land in the hands of small-scale farms.

The principle crop produced by small-scale farms was hay. Of the total number of small-scale farms in the country, over 44 percent produced hay in both 1982 and 1987. This represented 32 and 36 percent, respectively, of cropland harvested by these farms in 1982 and 1987. Corn was the second major small-scale farm crop with 27 and 24 percent, respectively, for 1982 and 1987. Soybeans were the third major crop, with 16 and 14 percent of the farmers growing this crop in 1982 and 1987, while wheat was the fourth major crop representing 14 and 11 percent, respectively. Of all small-scale farms in the country, vegetable producers represented about 3 percent in 1982 and 2 percent in 1987.

With respect to livestock, small-scale farmers primarily raised cattle, horses and ponies, chickens and hogs and pigs (Table 2.3). The most important livestock raised by many of these farmers were cattle and calves, primarily beef cattle. The second and third major enterprises for small-scale farm operators, respectively, were horses and hogs.

About 7 percent of small-scale farms in both 1982 and 1987 raised sheep and goats which was three times the number of small-scale farms raising vegetables in 1987 and about twice the number in 1982. Small-scale farms in the U.S., however, had generally more sheep and lambs than goats. The average small-scale farm had 53 head of sheep and lambs per farm in 1982 and 47 in 1987. Of all the livestock in this group of

TABLE 2.2: Major Crops Raised by Small-Scale Farms, 1982-1987

Crop	1982 Number	1982 %[a]	1982 Acres (000)	1987 Number	1987 %[a]	1987 Acres (000)
Corn	433,452	27.1	12,167 (1.2)[b]	364,453	24.2	10,576 (1.1)[b]
Hay	711,040	44.3	22,609 (2.3)	687,049	45.7	23,237 (2.4)
Oats	135,531	8.5	2,941 (0.3)	96,879	6.4	1,897 (0.2)
Orchards	96,101	6.0	1,114 (0.1)	92,390	6.1	994 (0.1)
Soybeans	253,588	16.0	12,326 (1.2)	216,214	14.4	10,345 (1.1)
Wheat	220,574	13.8	14,571 (1.5)	169,933	11.3	11,851 (1.2)
Tobacco	146,578	9.2	391 (0.0)	114,284	7.6	272 (0.03)
Vegetables	44,921	2.8	383 (0.0)	36,089	2.4	319 (0.03)
Other Crops	398,670	24.9	12,939 (1.3)	116,456	7.7	4,944 (0.5)

[a]Percent of all small-scale farms.
[b]Percent of total farmland in the U.S.

Source: Ibid.

TABLE 2.3: Major Livestock Raised by Small-Scale Farms, 1982-1987

Livestock	1982 Number (000)	1982 Head (000)	1982 Average Head Per Farm	1987 Number (000)	1987 Head (000)	1987 Average Head Per Farm
Cattle and calves	880 (54.9)[a]	16,582	18.8	741 (49.3)	14,730	19.9
beef cows	755 (47.1)	15,437	20.4	668 (44.4)	13,969	20.9
milk cows	125 (7.8)	1,145	9.2	73 (4.8)	761	10.4
Hogs and pigs	190 (11.8)	3,730	19.6	128 (8.5)	5,452	42.6
Sheep and lambs	75 (4.7)	3,962	52.8	69 (4.6)	3,221	46.7
Goats	38 (2.4)	674	17.7	39 (2.6)	850	21.8
Chickens[b]	191 (11.9)	16,660	87.2	126 (8.4)	7,500	59.5
Horses and ponies	325 (20.3)	1,597	4.9	333 (22.1)	1,804	5.4

[a]Percent of small-scale farms.
[b]Including Turkey.

Source: Ibid.

farms, chickens were averaging 87 and 59 head per farm for 1982 and 1987, respectively, while there were only 20 head per farm of cattle and calves for both periods.

Farm Income by Type of Farm

Disaggregating of income by receipts of the major enterprise components of the small-scale farmer helps us to understand the highly specialized nature of the country's small-scale agriculture. Table 2.4 shows that cash receipts from farming by small-scale operators vary significantly between crops and livestock. The trends in cash receipts, over time, are affected by the trends in land use, changing commodity prices, and yield per acre. This can be seen from the Table where receipts between 1978 and 1987 decreased by about 21 percent for livestock and 27 percent for crops. The decrease for the period, particularly between 1978 and 1982, reflects the strong recessionary trend that was observed in commodity price levels.

There were larger cash receipts by small-scale livestock producers in the country in 1987 than in 1982. The increase, which accounted for only 1 percent over 1982, was a result of higher livestock prices in the early and mid 1980s. Cash receipts in 1978 for crops were higher than receipts for livestock for the same time period. With the exception of 1987, the same trend has continued between crops and livestock as the source of cash receipts for the nations's small-scale farmers since 1978. Cash receipts from government payments accounted for less than 1 percent of the total between 1978 and 1982. The data for 1987, however, show an increase of 123 percent over 1982 due to disbursements of government funds for participants in the PIK program.

In addition to being balanced between livestock and crop enterprises, small-scale farming in the U.S. is diversified within each enterprise. Cattle (including calves) and dairy products are the largest contributors to cash receipts, representing a combined share of about 37 percent for 1978, 36 percent for 1982, and 42 percent for 1987 (Table 2.5). However, the mix of receipts between these two commodities has been changing

TABLE 2.4: Cash Receipts from Sales of Livestock, Crops, and Government Payments, 1978-1987

Item	1978	1982	1987
Crops ($000)	8,989,980	7,634,693	6,579,621
Percent of total	48.7	50.4	43.6
Livestock ($000)[a]	8,631,789	6,767,563	6,835,009
Percent of total	46.8	44.7	45.3
Government payments ($000)	820,000	742,000	1,658,177
Percent of total	4.4	4.9	11.0
Total	18,441,769	15,144,256	15,072,807

[a]Including poultry.

Source: Ibid.

TABLE 2.5: Source of Cash Receipts by Types of Farms, 1978-1987

Commodity	1978	1982	1987
	Percent of Total Receipts		
Livestock:			
Cattle and calves	29.1	30.6	36.9
Hogs	8.4	6.8	5.7
Dairy products	8.3	5.8	4.4
Poultry and poultry products	0.8	0.8	0.5
Sheep and sheep products	0.8	0.9	1.2
Other livestock	1.5	2.0	2.3
Crops:			
Grains[a]	31.5	33.4	29.5
Cotton and cottonseed	1.5	1.0	1.2
Hay	5.1	4.3	5.5
Tobacco	6.0	7.0	4.7
Vegetables	1.4	1.5	1.7
Fruits, nuts and berries	3.1	3.4	3.9
Nursery & greenhouse products	1.2	1.4	1.6
Other crops	1.2	1.0	1.0

[a]Includes: corn, wheat, soybeans, oats and sorghum.

Source: Ibid.

with dairy products' share declining continuously. Poultry and their products and hogs account for 9 percent of the country's small-scale farm receipts in 1978, with both commodities' share decreasing both in 1982 and 1987.

With respect to crops, grains lead the list of crops in contributing to cash receipts of small-scale crop farmers. Although prices for grains were below 1981 levels, cash receipts from these commodities in 1982 represented 33 percent of all cash receipts from crop production. This percentage receipt was close to that of 1978 and 1987 (Table 2.5). The constant importance of vegetables and an increasing share of receipts coming from tobacco (except in 1987) is the most significant trend that can be observed in Table 2.5. Compared to 1978 and 1982, hay was up in importance in 1987 and claimed over 5 percent of the cash receipts. Other crops not listed in the Table, including oats, rye and rice increased in importance in the late 1970s and early 1980s.

Receipts from crop sales are almost always higher than receipts from livestock sales. However, as shown in Table 2.4, in some years the reverse is true. This is true not only for the country as a whole but also for some states. For example, in 1986, livestock producers in Maryland showed a higher net farm income than crop producers. Of the livestock producers, dairy farmers reported the highest average net cash farm income of $21,494 per farm. Beef producers reported negative cash farm income averaging losses in excess of $3,000. Poultry and swine producers averaged about $16,000 income per farm.

On the other hand, crop farmers as a group reported net cash farm income averaging $14,000 per farm. Among crop farmers, the highest farm income was reported by tobacco producers, averaging $11,000, while vegetable farmers reported net cash farm income of $3,944, the lowest average positive value among crop producers (1).

General Characteristics and Trends for Farm Operators

Age of Operators

As indicated in Chapter 1, small-scale farming in the U.S. is a business operated by older, white males. In 1987, only 8 percent of the small-scale farm operators in the country were female. Of the total small-scale farmers in the country 97 percent are white and 1.5 percent are black with the other 1.5 percent representing American Indian, Asian, Spanish, and other races (2). As shown in Table 2.6, the average age of the nation's small-scale farmers for 1978, 1982 and 1987 was between 50 and 53 years. In 1987, more than 48 percent of these farmers were over the age of 55 years.

In the middle and late part of the 1970s, a relatively prosperous period of time in farming, there was a trend towards young farmers entering farming in the country. As a result, there were more farmers under 45 years of age in 1978 than in previous years. However, the trend was reversed in the early and mid 1980s (Table 2.6). In 1982 and 1987, there were fewer farmers under the age of 45 years than in 1978. In fact, farmers 65 years and older were the only group to show a slight decrease between 1978 and 1982 and an increase in 1987 in the older age group of farmers. This reversal in trend seems to be a result of the financial problem faced by farmers (the young farmer has been harder hit than the old farmer) in the nation.

Tenure of Operators

Tenure of farm operators in the U.S. is presented in Table 2.7. Tenure status deals with the rights of individuals in the use of land and other resources. The ways in which land, labor, and capital are combined under varying tenure arrangements influence the efficiency of production and the cost of farm products, thus affecting not only farm residents but all members of society (3).

TABLE 2.6: Farm Operators by Age, 1978-1987

Years	1978	1982	1987
Under 25	64,001	44,851	25,898
25 to 34	235,600	189,939	152,544
35 to 44	354,984	306,377	280,113
45 to 54	425,503	344,134	319,369
55 to 64	444,856	375,846	349,445
65 and over	365,331	343,022	376,982
Average age	50.2	51.2	53.1
55 and older (%)	42.9	44.8	48.3

Source: Ibid.

TABLE 2.7: Tenure of Operators and Form of Organization, 1978-1987

Item	1978 Number	1978 %	1982 Number	1982 %	1987 Number	1987 %
Tenure:						
Full owners[a]	1,266,170	67.0	1,144,042	69.4	1,054,931	70.1
Part owners[b]	397,309	21.0	322,423	20.1	302,043	20.1
Tenants[c]	226,776	12.2	167,704	10.4	147,477	9.8
Organization:						
Individual	1,719,905	91.0	1,453,426	90.6	1,363,342	90.1
Partnership	150,451	8.0	127,423	7.9	112,596	7.5
Corporation	13,318	0.7	15,426	1.0	19,860	1.3
Others[d]	6,601	0.3	7,894	0.5	8,653	0.6

[a] Farmers who only work on land they own.
[b] Farmers who work on land they own and also rent from others.
[c] Farmers who only work on land they rent from others.
[d] Includes cooperatives, estates or trusts, and institutions.

Source: Ibid.

Therefore, the tenure classifications shown in Table 2.7 are restricted to farm operators and their rights on the land they operate. As shown in the Table, full owners are those farmers who only operate land they own, part owners are those farmers who operate land which they own plus land which they rent from others, and those who only operate on land which they rented from others (or work on shares) are identified as tenants.

The majority of small-scale farmers in the country fully own and operate the land they farm. In 1978, about 67 percent of all small-scale farms were operated by full owners. However, their numbers decreased while their percentage of the total increased between 1978 and 1987. Part owners also decreased in number while their percentage as a portion of all farm operators since 1978 remained about the same. For example, their number decreased from 397,309 in 1978 to 302,043 in 1987, representing a decrease of 24 percent for the period. On the other hand, tenant farmers, with the exception of 1978, have remained about the same both in relative numbers and in percentages since 1978.

The constant percentage change in part owners of small-scale farmers as a proportion of total small-scale farm operators for the period reflects the overall interest in small part time farming in the country. Furthermore, its increase reflects the rise in the number of part owner operators, relative to other tenure status, who combine the security of an owned unit with economics of size provided by rental units to obtain a viable operation. According to the U.S. Senate Committee on Agriculture (4), the improved performance of farm machinery and other modern technology for operations on a large scale has given the modern, part-time farmer the capacity to expand and thereby increase his/her total net farm income.

Forms of Business Organizations

Small-scale farms are organized, as are all other businesses in the country, under four major types of business organizations. They are individually owned (sole proprietorship), partnership, corporations, and other types of ownership which include cooperatives, estates and trusts, and institutional ownership. Many of the farms owned by individuals or partnerships are basically family farms. Therefore, over 97 percent of the nation's small-scale farms were family farms in 1978; 1982; and 1987, respectively (Table 2.7).

Corporate and other types of small-scale farm organizations have been increasing over the years in the country. The percentage of these organizations, in aggregate, increased from 1 percent in 1978 to 1.5 percent in 1982, and to about 2 percent in 1987. Corporate farming has primarily concentrated on large livestock and specialty crop farms. The trend of increased corporate farming in the nation's small-scale farms indicates that, due to increasing capital and other resource needs in farming, this type of farming may be the way of future small-scale agriculture in the U.S.

Occupation and Residence

Full time small-scale farmers in 1978 represented about 42 percent of all small-scale farmers in the U.S. In both 1982 and 1987, 41 percent of the small-scale farmers reported farming as their principle occupation, representing a decrease of 1 percent in each period from what it was in 1978 (Table 2.8). Among the group of small-scale farmers who reported having an occupation other than farming in 1982, about 53 percent were between the ages of 35 and 54. This represented 89 percent of all the nation's farmers in this age group reporting other occupations.

In the past three census periods (1978; 1982; and 1987), over 145,000 small-scale farmers in the country worked between 100 and 200 days off the farm (Table 2.8). The percentage of those farmers who worked off the farm for 200 days or more remained about the same between 1982 and 1987. Off-farm work includes only that of the farm operator--not work of other farm family members. For those who are holding full time off-farm jobs, farming is becoming a part time job.

TABLE 2.8: Principal Occupation, Residence and Off-Farm Employment, 1978-1987

Item	1978	1982	1987
Place of Residence:			
On-farm	1,326,236	1,097,035	1,039,688
	(72.2)	(68.4)	(69.1)
Off-farm	359,471	337,980	344,103
Not reported	151,293	169,154	120,660
Principal Occupation:			
Farming[a]	796,284	660,045	621,665
	(42.1)	(41.1)	(41.3)
Other[b]	1,093,991	944,124	882,786
Days worked off-farm:			
None	608,416	467,997	447,883
One or more days	1,202,088	1,015,925	943,887
1-99 days	179,185	141,060	126,652
100-199 days	170,786	159,971	146,300
200 or more days	852,117	714,894	671,035
Not reported	9,771	120,247	85,681
Percent of 200 days	(45.1)	(44.6)	(44.6)

[a]Operator spent 50% or more of his/her time working in farming or ranching.
[b]Operator spent more than 50% of his/her time in occupation other than farming.

Source: Ibid.

With respect to residence, the number of farmers residing on the farm they operate has decreased slightly over the years, which is consistent with the decrease in the number of small-scale farmers. About 69 percent of farmers lived on their farms in 1987 compared to 72 percent in 1978. On the other hand, the number of farmers residing off the farm increased by two percent between 1982 and 1987. This, therefore, indicates that fewer small-scale farmers in the U.S. live on their farms while holding part-time or full time jobs off the farm.

Size Distribution of Farms

Although more small-scale farmers in the U.S. are becoming part time farmers, these farmers farm less land than the full time farmers. This is reflected in the growing percentage of farms less than 100 acres in size. In 1978, these small-scale farmers accounted for 54 percent of all small-scale farms in the country, but by 1987, they increased to 55 percent (Table 2.9). The percentage of land owned by this group of farmers has also shown a slight increase between 1978 and 1987.

Small-scale farms in the 100 to 179 and 180 to 499 acres size decreased in number between 1978 and 1987. For example, farms in the 100 to 179 acres size decreased by about 18 percent between 1978 and 1982, and seven percent between 1982 and 1987. This group of farms has also shown a decrease in acres of land that they farm. However, of all the small-scale farm sizes in the country, there has been an increase in number of farms in the 1,000 to 1,999 acre size. These farms increased from 22,812 in 1982 to 23,945 in 1987. The increase was also true in the acres of land they farm. For the same time period, this trend was also true for those small-scale farms in the category of 500-999 and 2,000 or more acre size.

TABLE 2.9: Farms by Size Distribution, 1978-1987

	1978		1982		1987	
Acres	Number	%	Number	%	Number	%
1 - 99	1,016,379	53.8	912,302	56.9	833,417	55.4
100 - 179	377,294	20.0	308,776	19.2	285,387	19.0
180 - 499	367,064	19.4	285,585	17.8	284,396	18.9
500 - 999	84,130	4.4	63,358	3.9	65,920	4.4
1000 - 1999	31,997	1.7	22,812	1.4	23,945	1.6
2000 or more	13,411	0.7	11,336	0.7	11,386	0.8

Source: Ibid.

Comparison of Small- and Large-Scale Farms

Enterprises and Farm Land Use

In 1987, about 72 percent of all farms in the U.S had sales of less than $40,000. These farms averaged 198 acres per farm and of these acres, 42 were in harvested cropland. These small-scale farms averaged $8,919 per farm of agricultural product sales, and they accounted for only 10 percent of all farm product sales for the same time period in the country. Crop sales contributed 49 percent of these sales, while the remaining came from livestock and poultry sales (Table 2.10).

On the other hand, only 28 percent of farms in the nation fell into the large-scale farm category. Obviously, they tend to be large-scale operations averaging 1,143 acres per farm, with an average value of agricultural products sales of $210,351 per farm. Of the 1,098 acres, 400 were harvested cropland. The majority of the agricultural product revenues (47%) of these farms came from the sale of livestock and livestock products.

Comparisons of enterprises and land use between these two sales classes show that there are large differences in cropping patterns. As shown in Table 2.10, small-scale farms harvested 48 percent of their cropland while 70 percent of the cropland of large-scale farms were harvested. The principal crop produced by small-scale family farms is hay, while large-scale farms tend toward other cash products such as corn and soybeans.

Cattle and calves were the primary livestock raised on both classes of farms. But the average herd size was 39 for small-scale farms, and 180 for large-scale farms. Small- and large-large farms each, respectively, held 36 and 64 percent of all the livestock inventory in the country. However, livestock sales as a percentage of total sales was 50 percent for small-scale farms compared to 47 percent for large-scale farms.

Although both sales classes raised livestock, small-scale farms raised more beef cattle, while large-scale farms included a substantial number of milk cows. Small-scale farms raised 44 percent of the nation's beef cattle while large-scale farms

TABLE 2.10: Selected Characteristics of Small- and Large-Scale Farms, 1987

		Average Per Farm	
Characteristics	Unit	Small	Large
Land in farms	Acres	197.8	1,143.0
Total cropland	do.	87.8	533.6
Cropland harvested	do.	42.5	374.2
Livestock inventory	Head	39.1	180.0
Poultry inventory	do.	5.0	634.0
Value of agricultural products	Dollars	8,919.5	210,350.7
Livestock and livestock products	do.	4,498.5	98,788.0
Poultry and poultry products	do.	46.0	21,765.0
Crop and crop products	do.	4,375.0	89,796.7
Value of land and buildings	do.	149,937.2	649,506.1
Owned	Acres	151.3	661.7
Rented or leased from others	do.	63.0	539.9
Rented or leased to others	do.	22.4	43.1
Value of machinery and equipment	Dollars	19,617.2	96,564.5
Cars and trucks	Number	1.3	2.6
Farm machines	do.	1.4	7.9
Farm production expenses	Dollars	9,163.9	161,844.9
Energy	do.	1,048.6	10,163.4
Livestock	do.	1,660.3	61,768.3
Crop	do.	1,570.7	21,275.5
Labor	do.	480.6	20,559.9
Interest	do.	1,040.1	11,310.3
Value of land and buildings (per acre)	do.	757.2	568.2
Value of machinery and equipment (per acre)	do.	99.1	84.5

TABLE 2.10 (Cont.)

Characteristics	Unit	Average Per Farm: Small	Large
Value of product sales per acre	Dollars	45.0	184.0
Production expenses per acre	do.	46.3	141.6
Crop sales per acre of land in farm	do.	22.1	78.6
Crop sales per acre of cropland harvested	do.	102.9	240.0
Value of agricultural products sales per dollar of asset value	Cents	0.05	0.3
Value of land and buildings per dollar of sales	Dollars	16.8	3.1
Value of machinery and equipment per dollar of sales	do.	2.2	0.5
Total production expenses per dollar of sales	do.	1.0	0.8
Cropland harvested as percentage of cropland	Percent	48.4	70.1
Cropland harvested as percentage of land in farms	do.	21.5	32.7
Crop sales as percentage of total sales	do.	49.0	42.7
Livestock sales as percentage of total sales	do.	50.4	47.0

Source: Ibid.

raised 56 percent. In contrast, however, large-scale farms raised 93 percent of all dairy cows in the country, while small-scale farms raised only seven percent.

As clearly shown in Table 2.10, the intensity of farm resource use differs significantly between small- and large-scale farms. Overall, farm resources are used intensively on large-scale farms, while they are used less intensively by small-scale farms. This implies that there is the potential for small-scale farms in the U.S. to engage in more intensive agriculture, given appropriate or relevant technologies[2] to farm their way out of their many problems, and to meet the needs of the world for food and fiber.

Farm Physical Assets

Small-scale farms in the U.S. accounted for 37 percent of the value of all land and buildings in the country, an average of $149,937 per farm. The aggregate value of farm machinery and equipment on small-scale family farms was also 34 percent of the value of all machinery and equipment in the nation. Their holdings of cars and trucks and farm machines were relatively high accounting for about two machines per farm. Of all the physical assets of this group of farms, the value of land and buildings represented 88 percent, while machinery and equipment accounted for only 11 percent (Table 2.10).

As will be expected, large-scale farms have the highest value of land, building, machinery and equipment. The value of land and buildings accounted for 63 percent of all the value of land and buildings in the country. This represented an average of $649,506 per farm. In the aggregate, large-scale farms in the country accounted for about 66 percent of all the values of machinery and equipment, averaging $96,564 per farm. A relatively higher percentage of the large-scale farms reported having cars and trucks, and farm machines per farm compared to small-scale farms. This is to be expected since large-scale farm operators have large sized farms on the average, which require larger inputs of capital assets.

A closer look at the physical asset per acre, however, shows a different situation for each sales class of farms. For example, the value of assets of land and buildings owned per acre was higher for small-scale farms than large-scale farms (Table 2.10). Furthermore, these farm operators had also a much higher asset value of land and buildings, machinery and equipment per dollar sales. The overall indication of the above results of the small-scale farms is that there is an under-utilization of resources by this group of farm operators in the U.S.

Farm Production Expenses

The total operating expenses of small-scale farms represented 13 percent of all farm expenses in the nation. However, it was higher than the value of all agricultural products sold by this group of farms. The largest portion of production expenses of these farms was for livestock, which run about 18 percent of the total expenses incurred by these farms. Crops and energy expenses accounted for about 17 percent and 11 percent, respectively, while interest accounted for 11 percent. With respect to expenses per dollar of sales, there was no significant difference between expenses per dollar of sales of small-scale farms and large-scale farms (Table 2.10).

Large-scale farms, as should be expected, showed the highest production expenses, averaging $161,845 per farm. Their expenses represented 87 percent of all production expenses in the country, and 77 percent of their farm product sales. The largest expenses of large-scale farmers was also for livestock. However, the expense was much lower than the revenue from sales of livestock and livestock products. The second and third largest expenses of these farms were respectively, crops and labor.

Notes

1. Crops raised depended primarily upon geographic location of the farm.

2. Relevant technology for the purpose of this book is defined as an agricultural practice which utilizes existing resources on the farm and fits the existing operating pattern of the farm.

References

1. *1986 Maryland Farm Finance Survey.* Maryland State Department of Agriculture, Agricultural Statistics Service, 1986.
2. *1987 Census of Agriculture for the U.S.* United States Department of Commerce, Bureau of the Census, Washington, D.C.
3. Maier, F. H., et al. *The Tenure Status of Farmworkers in the United States.* ARS/AMS/USDA, TB-1217, 1960.
4. U.S. Senate, Committee on Agriculture, Nutrition and Forestry. *Status of the Family Farm.* Ninety-sixth Congress, lst. Session, 1979.

3

Characteristics and Composition of Black Farm Operators and Their Farms

The Census of Agriculture data in Chapter 1 showed a substantial decrease in the number of farm operators and farmlands (structural change) in the U.S. in this century. The decline in number is generally true for both white and black farm operators, but the decline has been more significant for black farm operators.

However, while a great deal of attention has been given to the overall question of structural change in U.S. agriculture (representing 97 percent white operated farms) relatively little emphasis has been placed on the black farm operator. Several factors may account for this situation. The black farm operator mainly lacks the political influence, educational background and economic resources to demand that attention be paid to his problems.

The purpose of this chapter is, therefore, to identify the characteristics of black operated farms, compare and contrast resource use, and examine the current situation of black farm operators in the U.S. The chapter further discusses the history and importance of historically black Land-Grant institutions in the survival and development of blacks in general and black farmers in particular.

Characteristics of Black-Operated Farms

Number of Farms

The number of farms operated by blacks and whites declined since the beginning of this century. The general decline as previously indicated, however, has been much higher for the black farm operators than for whites.[1] For example, of the total 5,716,846 farms in the U.S. in 1900, blacks represented 746,717 (13 percent). Currently, there are less than 23,000 farms operated by blacks, representing about one percent of all farms in the country today, and a 97 percent decline from the number in 1900 (Table 3.1).

In contrast, in 1900, there were 4,970,129 white operated farms, accounting for 87 percent of the country's farm operators. In 1987, their number declined to 2,025,643, representing a loss of 59 percent over what it was in 1900. In other words, in 1987, there were 41 percent as many white farmers as there had been in 1900, while there were only three percent as many black farmers as there had been in 1900.

Over half of the loss of the nations's black farm operators occurred between the 1920s and the 1960s. At their peak number in 1920, there were 925,710 black farmers, comprising 14 percent of all farm operators. By 1964, less than 20 percent (184,004) of that number remained, representing a loss of 80 percent since 1920. Furthermore, in 1920 one American farmer in seven was black. By 1964, the ratio was 1 black farmer for every 17 farmers, and by 1982, this ratio had reached 66 to 1. Just in the last 19 years (between 1969 and 1987), black operated farms in the country declined at an average rate of 27 percent per census period. If the same rate of decline continues, 30 years from now their will be no farms that are operated by blacks in this country.

The declining trend in the number of farm operators in the U.S. has also been observed for other minority (American Indians, Asian, Hispanics and other races) farm operators. The data for the census years of 1974; 1978; 1982; and 1987, show a significant decrease for some minority farm operators.

TABLE 3.1: Farms Operated by Blacks and Whites in the U.S., 1900-1987

	Operators			
Year	White	Percent Change	Black	Percent Change
1900	4,970,129	--	746,717	--
1910	5,440,619	9.5	893,377[a]	19.6
1920	5,499,707	1.1	925,710	3.6
1930	5,373,703	-2.3	882,852	-4.6
1940	5,378,913	0.1	681,790	-22.8
1950	4,802,520	-10.7	559,980	-17.9
1954	4,298,766	-10.5	467,656[b]	-16.5
1959	3,419,672	-20.4	272,541	-41.7
1964	2,957,905	-13.5	184,004	-32.5
1969	2,626,403	-11.2	87,393	-52.5
1974	2,254,642	-14.1	45,594	-47.8
1978	2,182,215	-3.2	37,351	-18.1
1982	2,170,426	-0.5	33,250	-11.0
1987	2,025,643	-7.0	22,954	-31.0

[a] Alaska not included.
[b] Alaska and Hawaii not included.

Source: *Census of Agriculture for U.S.* 1964; 1974; 1982; and 1987, United States Department of Commerce, Bureau of the Census, Washington, D.C.

Therefore, it is safe to conclude that if the declining trend in the number of farms operated by blacks and other minorities continues, there will be few (may be none) of these farm operators in the country by the end of the first decade in the year 2000.

Amount of Land In Farms

Since 1900, the amount of farmland owned by black farm operators has steadily decreased in the U.S. For example, in 1959 there were 1,124 million acres of farmland in the country. Of this, blacks owned only 51 million acres (Table 3.2). Between 1959 and 1987, this total had decreased to less than 3 million acres, representing a loss of 95 percent in just 28 years. In other words, black farmland owners accounted for 4.5 percent of the total land in farms in 1959 compared to 95 percent for whites. However, by 1987, black farmland owners accounted for only 0.3 percent of the total farmland, while white farmland ownership accounted for 94 percent of the total farmland, a net loss of only one percent for white farmland owners compared to a 4.2 percent for blacks.

Although the total acres of farmland has decreased over the years, the average size of farms in the country has significantly increased (Table 3.2). In 1959, the average acre size per farm was 303, but by 1987, the average size increased to 462. This increase in size between 1959 and 1987 represented about a two percent annual growth rate for the country as a whole. If we assume that this annual growth rate will continue, it is safe to suggest that the average size will increase and reach 510 acres by 1990 and 598 acres by the year 2000. Given the decline in the number of farms with a faster rate than land in farms, this result is what would be expected.

However, the average farmland size owned by black farmers in the U.S. was, and continues to be, proportionately less than the average size of farmland owned by white operators. In fact, the acreage size of black owned farms was smaller than the average size for the country. The census data in Table 3.2 shows that the average size of black farms was more than four

TABLE 3.2: Number of Farms, Acres of Land in Farms, and Average Size of Farms by Race, 1959-1987

	Year						
Item	1959	1964	1969	1974	1978	1982	1987
All Farms:							
Number of farms (000)	3,711	3,158	2,730	2,314	2,258	2,241	2,088
Land in farms (million acres)	1,124	1,110	1,061	1,017	1,015	987	964
Average size (acres)	303	352	389	440	449	440	462
White:							
Number of farms (000)	3,420	2,958	2,626	2,255	2,182	2,171	2,026
Land in farms (million acres)	1,070	1,063	1,055	1,008	950	925	904
Average size (acres)	313	359	402	447	435	426	446
Black:							
Number of farms (000)	273	184	87	46	37	33	23
Land in farms (million acres)	51	10	8	9	4	3	2.6
Average size (acres)	187	57	87	194	108	105	115

Source: Various issues of the *Census of Agriculture for U.S.*, United States Department of Commerce, Bureau of the Census, Washington, D.C.

times less than the average size for the nation for the period between 1964 and 1987. However, the average size of white farms was consistently higher than the average size for the country between 1959 and 1974 while it has been slightly below the national level since 1974.

Several factors contributed to the huge decline in the number of farms and amount of land in farms of black operators over the years. Some examples are the abandonment of the tenant system of crop production (such as tobacco and cotton); the marginal economic situation of black farmers, and the consequent difficulty in sustaining a farm operation; the smaller acreage of land and sales from black operated farms;[2] and the failure of young blacks to enter farming (i.e., the problem of intergeneration transfer of property). In addition, the revolution in cotton productions during the decades between 1920s and 1960s due to machines and chemicals, resulted in rural black migration from the farm to the cities.

Types of Farms

Farmers in the U.S. have traditionally produced tobacco, corn and other cash grains as well as livestock. As shown in Table 3.3, livestock represented the largest proportion of black and white operated farms in 1987. However, the second and third primary enterprises for black operated farms were cash grain and tobacco, while it was cash grain and other field crops, respectively, for whites. Poultry farm operation ranked seventh for white operators, while it was fruits and nuts for black operators.

Vegetable farms did not rank in the top three of farm enterprises for either of the two group of farm operators; however, the percentage was higher for blacks than for whites. On the average about 3 percent of black operated farms were classified as vegetable farms. This result suggests that black as well as white small-scale farm operators in the country have the potential to shift their farm operations from tobacco and other traditional enterprises to vegetable production to meet the

TABLE 3.3: Type of Farms by Race, 1987

	White		Black	
Type of Farm	Number	Percent	Number	Percent
Cash grain	453,014	22.4	3,150	13.7
Tobacco	84,904	4.2	2,460	10.7
Vegetables	25,618	1.3	703	3.1
Fruits and nuts	80,789	4.0	298	1.3
Other field crops[a]	179,269	8.8	1,961	8.5
Poultry and eggs	37,755	1.9	285	1.2
Livestock[b]	1,086,280	53.6	13,207	57.5
General	78,014	3.8	890	3.9
Total	2,025,643	100	22,954	100

[a]Includes horticulture.
[b]Includes dairy and animal specialties.

Source: *Census of Agriculture for U.S.*, 1987, United States Department of Commerce, Bureau of the Census, Washington, D.C.

changing economy and market signals in agriculture in the nation. Such a shift will undoubtedly help improve their income from farming.

Black farmers have been, and continue to be, relatively slow to adopt to changes that are taking place in the agricultural economy in the past. Various factors were, and still are, responsible for this situation. Among them are lack of capital, unavailability of information, small-scale operations, and unwillingness to break traditional production processes.

Farm Product Sales

Total farm product sales of the nation have increased over the years (27 percent between 1978 and 1987), reflecting a strong inflationary trend in commodity prices. Of the total cash receipts in 1987, for example, over 97 percent of the share was received by white farm operators, while only 0.2 percent was received by blacks. Of the total receipts of agricultural products for blacks, 45 percent came from sales of livestock and their products, while the remaining 55 percent came from sales of crops and their products. For whites, the percentage was 57 and 43, respectively (Table 3.4).

According to the 1987 census data in Table 3.4, 81 percent of all black farm operators were in the lower sales (less than $10,000) category, 8.5 percent in the $10,000 to $19,999 sales category and about 11 percent in the more than $20,000 category. By contrast, white farmers are more equally distributed across sales category, 23 percent in less than $2,500 category, 26 percent in the $2,500 to $9,999 category, 12 percent in the $10,000 to $19,999 category and 39 percent in the over $20,000 category.

About 89 percent of black farmers have sales less than $20,000. As a result of this and other factors, net household income of black operators has been, and continues to be, below that of non-metropolitan median household income. This is to say that black farmers in the country are low income farmers

TABLE 3.4: Value of Agricultural Products and Farms by Value of Sales, 1987

Item	Unit	White	Percent	Black	Percent
Farm Production[a]:					
Value of agricultural products	($ mill.)	132,707	97.5	332	0.2
Value of agricultural products per acre	($)	147	-	126	-
Value of livestock and livestock products[b]	($ mill.)	76,259	98.9	150	0.2
Value of crops and crop products[c]	($ mill.)	56,447	95.8	182	0.3
Farms by value of sales[d]:					
Less than $2,500	Number	466,848	23.0	10,662	46.4
$2,500 to $9,999	do.	519,240	25.6	7,866	34.3
$10,000 to $19,999	do.	244,466	12.1	1,943	8.5
$20,000 or more	do.	795,089	39.2	2,483	10.8

[a]Percent for this section is of U.S. total.
[b]Includes poultry and their products.
[c]Includes nursery and greenhouse products.
[d]Percent for this section is only for the group.

Source: Ibid.

and yet over half of them report farming as their main occupation.

This result, therefore, suggests that income from farming for blacks is disproportionate to the time spent in farming. Furthermore, it suggests that black farmers have room to improve their productive efficiency and increase their agricultural income. Most of these farmers can become more productive with adequate management techniques, removal of institutional barriers, and credit financing.

Characteristics of Black Farm Operators

Age of Operators

Farming in the U.S., as indicated in the previous chapters, is an occupation dominated by older, white and male farmers. For example, in 1978; 1982; and 1987 they represented 82; 90; and 93 percent, respectively, of all farm operators in the country. Among the white farm operators, about 45 percent were 55 years old or older in 1987, while it was 61 percent for black farm operators (Table 3.5).

Proportionately more black farm operators are found in the upper age brackets. Of all the black farm operators in 1987, 24 percent were between 55 and 64 years old and 36 percent over 65 years old, compared to 24 and 21 percent, respectively, for white farm operators. In short, more black farm operators were in the upper age brackets (55 years or more) and fewer were younger (less than 34 years) than white farmers in 1987, which has been a trend for many years for this group of farmers. In contrast, an increasing trend for white younger farmers was developing during the time it was declining for black farmers, indicating that more young white men are entering into farming than young black men.

Table 3.5 also shows the average age of white operators to be lower than that of black operators, even though the average age of both groups was more than 50 years old. Due to age, as well as health, and lack of training, many of these older black

TABLE 3.5: Farm Operators by Age, 1987

Age of Operator	White		Black	
	Number	%	Number	%
Under 25	35,221	1.7	172	0.7
25 to 34	237,355	11.7	1,261	5.5
35 to 44	399,603	19.7	3,309	14.4
45 to 54	441,487	21.8	4,285	18.7
55 to 64	480,658	23.7	5,574	24.3
65 and over	431,319	21.3	8,353	36.4
Total	2,025,643	100	22,954	100
Average age	52.0 years		57.9 years	

Source: Ibid.

farmers are, and will continue to be, occupationally and geographically immobile. This, therefore, suggests that any strategy of programs to improve the well-being of black farm operators should take into consideration the difference in needs between younger and older operators.

Tenure of Operators

Tenure operator categories by race are presented in Table 3.6, and these tenure characteristics of farmers are a primary measure of resource control in agriculture. Between 1959 and 1987, more than half of the nation's farm operators were full owners. Of all farm operators, part owners represented, on the average, about 26 percent, while tenant farmers represented over 14 percent for the same time period.

Black full owners in 1959 accounted for only 4.6 percent of all full owners of farms in the U.S., while the remaining 95 percent represented white full owners. By 1987, the percentage for blacks had decreased to about one percent, while it increased to 97 percent for whites. Among black farm operators, however, full owners in 1959 represented about 35 percent of all black farm owners, while white full owners accounted for about 59 percent of all white farm operators in the nation. In 1987, the percentage of full owners among black farmers increased over what it was in 1959, while it was about the same for full owners among white operators.

Typically, black farmers made up a larger percent of the tenant operated farms than did white farmers in that category. In 1959, 51 percent of the black farmers were tenants compared to 18 percent for white operators. Although the percentage of black tenant farmers remained slightly above that of whites, by 1987 it had declined by more than six-fold over what it was in 1959. Over the years, relatively more black farmers than white farmers represented tenant operated farms, because they were, for various reasons, precluded from ownership of land, particularly in the South. One of the reasons was economic, such as price competition for farmland, limited collateral, and lack of credit.

TABLE 3.6: Tenure of Operators by Race, 1959-1987

Tenure of Operator	Year 1959	1964	1969	1974	1978	1982	1987
All Farms:							
Full-owners (000)	2,117	1,818	1,706	1,424	1,298	1,326	1,239
	(57.1)[a]	(57.6)	(62.5)	(61.5)	(57.5)	(59.2)	(59.3)
Part-owners (000)	835	782	672	628	681	656	609
	(22.5)	(24.8)	(24.6)	(27.1)	(30.2)	(29.3)	(29.2)
Tenants (000)	757	558	353	262	279	259	240
	(20.4)	(17.7)	(12.9)	(11.3)	(12.3)	(11.6)	(11.5)
White:							
Full-owners (000)	2,017	1,740	1,650	1,385	1,401	1,292	1,199
	(58.9)	(58.8)	(62.5)	(61.4)	(58.6)	(59.1)	(59.2)
Part-owners (000)	792	747	655	616	695	643	595
	(23.2)	(25.3)	(24.8)	(27.3)	(29.0)	(29.4)	(29.4)
Tenants (000)	613	471	334	254	300	252	232
	(18.0)	(15.9)	(13.0)	(11.1)	(12.5)	(11.5)	(11.4)
Black:							
Full-owners (000)	97	79	55	39	22	21	15
	(34.8)	(39.3)	(61.2)	(65.3)	(59.1)	(62.2)	(65.2)
Part-owners (000)	41	35	16	12	10	9	6
	(14.6)	(17.4)	(17.9)	(20.7)	(27.8)	(26.2)	(26.1)
Tenants (000)	142	87	19	8	5	4	2
	(50.6)	(43.3)	(20.8)	(13.9)	(13.1)	(11.3)	(8.7)

[a]Figures in parentheses indicate group percentages.

Source: *Census of Agriculture for U.S.*, 1969, 1982. United States Department of Commerce, Bureau of the Census, Washington, D.C.

With respect to part ownership, black and white part owners, respectively, represented about 26 percent and 29 percent of all black and white operated farmlands in 1987. These percentages are relatively higher than what they were in the 1950s; 1960s; and 1970s (Table 3.6). These data suggest that as the number of farms declined and the average size increased, the tenure of farm operations, over the years, has also changed. In fact, in the past two decades the trend in the U.S. has been toward increased leasing of land for agricultural production through an increased number of acres operated by part owners. As a result, part owners have increased in number as a portion of all farm operators, at least since 1959. On the other hand, full owners and tenants have declined in relative numbers since then.

The concentration of blacks in the lower economic classes (small farm) of farm operators in the U.S., observed in Table 3.4, is closely related to patterns of operator characteristics, tenure and type of farm that were discussed above. This, however, is not to say that because one is a certain type of farmer a lower income would necessarily be expected. It simply means that, upon grouping of farm operators by race as well as economic class, and inspecting the general measures of their characteristics, one can observe certain patterns.

Off-Farm Work and Principal Occupation

For many years farming has been the primary source of family income of farmers in the country. This was particularly true for black farm operators more than for whites. For example, in 1987, farming was the principal occupation for about 44 percent of the black farmers in the U.S. (Table 3.7). Many of the farmers in the country also held jobs in an occupation other than farming. Among the black farm operators in 1987, 56 percent reported having an occupation other than farming, compared to 45 percent for white farm operators. As a result, off-farm income has represented an important alternative source of income for farm families, particularly for small and limited resource black farmers.

TABLE 3.7: Farm Operators by Principal Occupation, Residence, Off-Farm Work and Organization, 1987

Characteristics	White Number	White Percent	Black Number	Black Percent
Off-Farm Employment:				
None	822,531	40.6	8,688	37.8
Under 100 days	192,650	9.5	2,426	10.6
100 to 199	171,278	8.5	2,348	10.2
200 days or more	715,550	35.3	7,559	32.9
Not reported	123,634	6.1	1,953	8.5
Total	2,025,643	100	22,954	100
Operator Residence:				
On-farm operated	1,451,144	71.6	13,255	57.7
Not on-farm operated	422,568	20.9	6,468	28.2
Not reported	151,931	7.5	3,231	14.1
Total	2,025,643	100	22,954	100
Form of Organization:				
Sole proprietorship	1,755,395	86.7	20,981	91.5
Partnership	194,411	9.6	1,508	6.6
Corporation	64,798	3.2	240	1.0
Other[a]	11,039	0.5	225	1.0
Total	2,025,643	100	22,954	100
Principle Occupation:				
Farming	1,108,805	54.7	10,071	43.9
Other	916,838	45.3	12,883	56.1
Total	2,025,643	100	22,954	100

[a]This category includes cooperatives, estates and trusts, and institutional ownership.

Source: *Census of Agriculture for U.S.*, 1987, United States Department of Commerce, Bureau of the Census, Washington, D.C.

The trend of off-farm employment in the nation has varied over the years for both black and white farm operators. In 1959 about 42 percent of the black farm operators and 48 percent of the white farm operators were working off the farm. By 1987, the percentage of both groups reporting off-farm work had increased, reaching 53 percent for white and 54 percent for black farm operators.

Although the percentage of farm operators reporting any off-farm work is about the same for both black and white farm operators in the country in 1987, there was a difference in the number of days worked. With the exception of those who worked 200 days or more, a large percentage of black farmers reported working more days off the farm than did whites. For example, among all black farmers reporting off-farm work, about 10 percent represented those who worked between 100 and 199 days off-farm work per year, while it was 8 percent for whites. A higher percentage of black farmers were also found in the category reporting under 100 days off farm work than white farmers. On the other hand, a lower percentage of black farm operators were found in the category reporting no off-farm work than white farm operators.

The relationship between off-farm employment or other occupation and farming is particularly important for analyzing farm household income, employment, and agricultural policy initiative for both group of farmers in general, but particularly for black operators. In fact, given the high unemployment rate for blacks in the country, opportunities for future off-farm employment for blacks is questionable. Therefore, it is important to develop a strategy for a farm (rural) development program in the nation that should take into account the amount of labor that is available to the farming operation.

Types of Farm Organizations

As indicated in Chapter 2, many of the partnerships and individually owned farms in the U.S. are basically family organizations. As a result most of the nation's white and black small-scale farmers operate truly family farms (Table 3.7).

However, among the black farmers, sole proprietorship claimed about 91 percent with partnership taking about 7 percent of ownership. Corporate and other type organizations consisted of about 2 percent in 1987. Of all white farm operators in the country, individual ownership represented more than 84 percent, while partnership represented about 10 percent. Also in 1987, corporate and other ownership accounted for about 4 percent.

The organization and tenure trend of black farm operators demonstrates that this group of farmers has the resources to make them as economically responsive as other farmers in the nation, and as a group, they have the potential to increase food and fiber supply significantly. This in turn means an increase in income which will have a positive impact upon the well-being of black farmers and the rural economy in which they reside.

History and Importance of 1890 Institutions

Black farmers have lived and worked in a total environment of technological change, government policies and programs, and distorted racial attitudes that have kept them in a difficult position to survive relative to white farmers. The only institutions, with all their funding difficulties over the years, that have served the needs of all small-scale farmers and particularly black farm operators have been the 1890 (black) Land-Grant institutions. The purpose of this section is to discuss the history, role and contribution of the historically black Land-Grant institutions in the general development programs of blacks and black operated farms.

History

The Congress of the United States passed the first Morrill Act of 1862 which provided the establishment of Land-Grant institutions in each state (1). These institutions were established to educate citizens in the fields of agriculture, home economics, the mechanical arts, and other useful professions. In the South, where there was a high concentration of blacks, blacks were not permitted to attend the institutions first established under the Morrill Act of 1862.

Congress passed the second Morrill Act of 1890, primarily to increase appropriation to the already established 1862 (white) Land-Grant colleges; it also included the provision to establish colleges for blacks (2). These institutions were intended to serve as counterparts to the 1862 colleges, established by the First Morrill Act, in the southern states[3] that insisted on a separation of races. Thus, the black Land-Grant institutions are referred to as the "1890 institutions".

Although the Act contained a "separate but equal" facilities clause, no portion of the appropriation from the second Morrill Act could be used for the construction of new campuses. Only already established schools could be designated 1890 colleges or the states themselves could pay for construction of new colleges. As a result, only 17 black Land-Grant colleges, including Tuskegee University, remain today. In contrast, however, 1862 colleges had received sizable land grants from the federal government to establish their campuses (3).

Various legislation by the federal and state governments has denied the black Land-Grant institutions funds and support, to which they were entitled, to fulfill their basic mission of teaching, research and extension. For example, the Hatch Act of 1887 provided funds for the establishment of experimental station programs. However, the Act directed State legislatures in the sixteen states with 1862 and 1890 Land-Grant colleges to designate the college or colleges that would receive these federal funds. In all these states, the 1862 colleges have always been chosen. As a result, 1890 colleges received little or none of the funds supposedly provided by this Act (3).

Various research conducted by 1862 Land-Grant colleges and universities demonstrate how these colleges and universities primarily served the interests of white high-income farmers. Research on the mechanization of crops, for example, was responsible for increasing the minimum size of farms; this benefited mostly white, large-scale farm owners and resulted in the rapid decline of black family farmers in general and tenant farmers in particular (4,5). At best, the 1862 Land-Grant Universities research, even today, is geared to the needs of no more than 30 percent of the country's farmers, those with an annual income of $40,000 and over in 1987.

The tragedy of black farm operators continued with the enactment of the Smith-Lever Act of 1914, which established the Cooperative Extension Service of the USDA. The Act says that where a state has two Land-Grant colleges, the appropriation for extension work "shall be administered by such college or colleges as the state legislature...may direct." Senator Smith of Georgia, author of the act, made the purpose of this provision quite clear, "we do not...want the fund if it goes to any but the white colleges" (6). The purpose was accomplished by the state legislatures in each of the states with white and black Land-Grant colleges, by directing all federal extension funds to 1862 colleges with compliance by USDA (3).

In addition to the institutional racism in the funding of extension service at black colleges, the extension service of the white Land-Grant colleges in the South discriminated against blacks. Agents were segregated for extension training and the training of white agents was longer, more comprehensive, and more detailed than for blacks. In areas where there was a large number of black farm operators, there were no black extension personnel, and the white extension personnel who were assigned provided little assistance to black farmers (4,5,7).

A study for the Committee on Labor and Public Welfare of the U.S. Senate found that 80 percent of the black farmers in Alabama were never visited by extension agents (8). Another study by the office of the Inspector General of USDA, for the same committee, consistently found that extension services in the southern and border states discriminated persistently from 1965 to 1972 in such matters as the holding of segregated meetings, the use of segregated mailing lists, the refusal to permit white agents to serve blacks or black agents to serve whites, and discriminatory hiring and promotion policies (8). As a result of these discriminatory practices, survival for black farm operators, particularly in the South, were and continue to be more difficult than for whites.

The discriminatory funding and operational practices of the Land-Grant college system have persisted despite the passage of the 1964 Civil Rights Act. Title VI of this Act gives minorities equal protection, access to, and benefits from federally financed programs. However, black Land-Grant institutions have never been given full access to and benefits from these funds, until 1972. For example, in 1968 the USDA gave less than $400,000 to the 1890 Land-Grant colleges while it gave $60 million to the 1862 colleges in the same states, a figure 150 times more than that of the 1890 institutions (9).

Moreover, in 1971 the USDA, through the Hatch and Smith-Lever Acts alone, allocated $87 million to white Land-Grant colleges in states with black Land-Grant institutions. For the same year, however, funds totalling $283,000 were granted to the 1890 colleges in the same states under the PL 89-106 program—the main source of federal agriculture funding to the predominantly black institutions (8). In other words, in 1971, the white Land-Grant institutions in the sixteen states received approximately three times more from the two agricultural programs than the 1890 colleges in the same states received from all major federal programs combined. Of the total funds to the 1890 colleges, the allocation to each individual institution ranged from $12,000 to $22,000 or an average allocation of $17,687 per institution for the year.

In 1972, Congress finally took action to partially correct this inequality by increasing the USDA appropriation for the 1890 colleges to $8.6 million. However, for the same year, the 1862 universities received about $94 million, almost $85 million more than the appropriation to the 1890 institutions (8). In 1977, a formula funding law was enacted that will tie the funding level of the black Land-Grant universities to the funding level of the white Land-Grant universities (10).

Based on this formula funding, the annual allocation of funds for the 17 black Land-Grant institutions has increased over the years. For example, in 1985 the total fund from USDA granted to these institutions for research purposes was $23.5 million. Likewise, the formula fund for extension services granted to these institutions has also increased, ranging from $4 million in 1972 to about $17 million in 1985.

While research and extension service funds have increased since 1972, the 1890 institutions still receive much less money than that allocated to all colleges and universities--in violation of the intent and spirit of the Morrill Act of 1890. As a result, today we have "separate but unequal" black Land-Grant colleges and universities. These institutions still remain far behind in their ability to continue and strengthen programs important to the mission of these institutions--to impact on the problems of all people in general, and blacks in particular.

Importance

Despite years of neglect by federal and state programs and policies determined to keep black institutions separate but not equal to 1862 institutions, the 1890 institutions provided access to education, research and extension services for their clientele without regard to race, creed, religion or socio-economic conditions. Although they are committed to provide services to all people, these institutions principally cater to low-income and disadvantaged blacks, because it is this sector of the population that white Land-Grant institutions were unwilling to serve in the past, and are ill-equipped to serve today.

The 1890 colleges have made educational opportunities possible for blacks, assuring them universal access to higher education. As a result, they have been a valuable source for many of America's college trained black population which includes, agricultural scientists, architects, educators, engineers, lawyers, nurses, and veterinarians, just to mention a few. These professionals have rendered unlimited valuable services not only to their states but to the nation as a whole.

The 1890 colleges' success as institutions devoted to the goal of equal educational opportunity is also evident in their ability to retain low-income and disadvantaged students, black or white. Today, approximately 60,000 students who are enrolled in these institutions come mostly from low-income families. Studies have shown that the attrition rate of blacks, especially those who are financially or educationally disadvantaged, is higher in predominantly white colleges than in predominantly black colleges (11). Therefore, the 1890 institutions stand today as insurance against present barriers which would prevent many blacks from attaining a college education.

About 60 percent of the nation's blacks still live in the states with the 1890 colleges and universities and close to 50 percent of these people live in rural areas and have an estimated annual income of less than $5,000. Most of these people are small-scale and limited resource black farm operators. The research and extension programs of the 1890 institutions have provided services in the past, and continue to do so today, to overcome the special problems of this group of farmers.

Most of the funds at the 1890 colleges are used to aid the rural poor rather than displace farmers, as the 1862 Land-Grant colleges and universities do. For example, in 1972, the 1890 institutions received a total of $12.6 million for research and extension services. A great majority of these funds were used to assist people to improve their incomes, welfare, health and the development of their communities (8).

Furthermore, the work done by researchers, extension specialists, and educators at the historically black Land-Grant colleges and universities on small-scale agriculture has helped find new techniques that are not only useful in the U.S., but can be transferred to the developing world. The exchange of scientific information with the developing countries dates back to 1900s when Dr. George Washington Carver, the great scientist, sent a team of agricultural experts to Togo.

The policies and programs of state and federal governments provide funds to 1862 colleges and universities who spend disproportionately more of these funds to serve prosperous big farms, which are mostly white owned. On the other hand, the 1890 colleges receive disproportionately less money to serve low-income small-scale rural farmers, who are mostly black. This unequal funding of black Land-Grant colleges is ironic because these institutions serve the very people the Land-Grant college system was established to serve--the uneducated and poor farmer.

It is additionally saddening to hear statements these days from USDA praising the contributions made by the black Land-Grant universities to humanities by providing services to the hard-to-reach and limited resource farmers in general, blacks in particular (8). This is the very agency overseeing the unequal allocation of Land-Grant college appropriation.

In his recent memorial lecture given at Tuskegee University, the former Secretary of Agriculture, Richard E. Lyng, stated the long history of the 1890 Land-Grant colleges working with limited resource farmers. Furthermore, he praised black Land-Grant colleges for having rapport and a unique channel of communication with the unreached and hard to reach in rural areas of the South. He also suggested that with the slightly increased research and extension funding, these Land-Grant institutions have already begun to carry out innovative programs, particularly helping rural poor (12). Unfortunately, the important role 1890 colleges played, and continue to play, in the nations, educational, research and extension systems and society at large, has yet to be fully appreciated and compensated for in federal financial aid and educational policies.

Notes

1. Throughout this chapter the data for the nations's white farm operators is used for the purpose of comparison.

2. Almost all black operated farms in the U.S. are small-scale farms.

3. Theses states are the states listed in Table 3.3.

References

1. U. S. Congress. *Public Law 12 Stat. 503, Ch. 130,* 1862.
2. U. S. Congress. *Public Law 26 Stat. 417, Ch. 841,* 1890.
3. Williams, T. T. "Teaching, Research and Extension at Predominantly Black Land-Grant Institutions." in *Human Resources Development in Rural American: Myth or Reality.* Ed. by T. T. Williams, Human Resources Development Center, Tuskegee University, Tuskegee, Alabama, 1986.
4. Huffman, W. "Black-White Human Capital Differences: Impact on Agricultural Productivity in the U.S. South," *American Economic Review,* 71 (1981).
5. Christian, V. L., et al. "Agriculture," in *Employment of Blacks in the South.* Ed. by R. Marshall and V. L. Christian, Austin, Texas: University of Texas Press, 1978.
6. Schuck, P. "Black Land-Grant Colleges: Separate and Still Unequal," *Farmworkers in Rural America.* U.S. Congress, Senate Hearing before the sub-committee on Migratory Labor of the Committee on Labor and Public Welfare, 1971-1972.
7. Schor, J. "Black Presence in Cooperative Extension Service Since 1945: An American Quest for Service and Equity," in *Human Resources Development in Rural America: Myth or Reality.* Ed. by T. T. Williams, Human Resources Development Center, Tuskegee University, Tuskegee, Alabama, 1986.
8. U.S. Senate, Committee on Labor and Public Welfare. *Farmworkers in Rural America.* Ninety-second Congress, 2nd Session, 1972.

9. Payne, W. "The Negro Land-Grant Colleges," *Civil Rights Digest,* 3 (1970).
10. U.S. Congress, *Public Law 95-113, Section 1145,* 1977.
11. *Access of Black Americans to Higher Education.* National Advisory Committee on Black Higher Education and Black Colleges and Universities, United States Office of Education, 1979.
12. Lyng, R. E. "The Rural South: Looking to the Twenty-First Century," in *Ushering in the Twenty-First Century: Emphasis on the Rural South.* Ed. by T. T. Williams, Tuskegee University, Tuskegee, Alabama, 1987.

4

Financial Conditions and Trends for Farms and Farm Operators

The future outlook of American agriculture depends on the current and past financial situation of the country as a whole. Currently, a large number of the nation's 2.2 million farmers suffer from financial difficulties, with many of them facing serious survival problems. Related family problems--stress, increased divorce, and even suicide are also on the rise.

This current aggregate farm sector financial stress evolved from, among other reasons, lower commodity prices, lower farm exports, and lower farm land values. Lower commodity prices were a result of large supplies and weak demand while lower farm exports were due mainly to the world-wide recession and the strong U.S. dollar. Price of land, hence land value, declined reacting to diminished prospects for income growth and higher returns on alternative investments. Furthermore, the financial problems are further symptoms of the structural changes that are taking place in American agriculture.

The above information helps us better understand the general picture of the financial situation in the farm sector of the country and decisions that will be faced by farmers in the next few years--maybe even in the next decade. However, any realistic assessment of the future of agriculture in the U.S. should include the study of the farm financial picture in the past. And a look at the farm income trends of the nation provides a backdrop against which the current and future financial situation may be assessed.

Trends in Farm Income

Total operator farm income (gross farm income) for U.S. farmers increased between 1964 and 1984 and declined in 1986. For example, in 1984 the gross farm income of $168,507 million accounted for an increase of 291 percent since 1964 and over 33 percent since 1978 (Table 4.1). On the other hand, however, farm production expenses have also moved persistently higher than gross farm income for the same time period. The increase in farm production expenses in 1984 represented about 348 percent that of 1964 and 38 percent that of 1978.

The nation's average gross farm income has also increased over the years, averaging $12,474 per farm in 1964 and $73,605 in 1986, while production expenses for the same time period averaged $9,202 and $55,179 per farm, respectively. The increase in production expenses per farm between 1964 and 1986 was 10 percent higher than the increase in gross farm income per farm for the period (Table 4.1).

From mid to late 1960s and early 1970s, net farm income[1] increased significantly due to an increase in domestic prices and growth in export markets. In fact, the nation's net farm income actually more than doubled between 1964 and 1974 (Table 4.1). In 1978, net farm income decreased to $23,306 million, $5,570 million less than what it was in 1974 but increased to $24,908 million in 1982. These changes indicate the erratic nature of net farm income from one year to another. Although the changes in net farm income have been erratic, they have been increasing since 1982, reaching $40,756 million in 1986. As a result, in 1986 net farm income for the country averaged about $18,426 per farm (Table 4.1).

For all the nations's farmers, net farm income as a proportion of expenses, at 41 percent was higher in 1974 than any other time between 1964 and 1986. However, it declined since 1974, reaching 33 percent in 1986. One possible explanation of this situation is the inflationary trend the economy had experienced during this period. As a result there has been an increase in off-farm income.[2] For example off-

TABLE 4.1: Farm Income and Expenses of All Farms, 1964-1986

Item	Year					
	1964	1974	1978	1982	1984	1986
Gross farm income ($ mil)	43,121	99,856	126,555	164,886	168,507	162,808
Average/farm ($)	12,474	35,727	51,952	68,687	72,370	73,605
Farm production expenses ($ mil)	31,812	70,980	103,249	139,978	142,669	122,052
Average/farm ($)	9,202	25,395	42,385	58,311	61,273	55,179
Net farm income ($ mil)	11,309	28,876	23,306	24,908	25,838	40,756
Average/farm ($)	3,271	10,332	9,567	10,376	11,097	18,426
Off-farm income ($ mil)	11,637	28,135	29,703	36,428	38,318	44,708
Average/farm ($)	3,366	10,066	12,193	15,175	16,457	20,212
Net farm income as % of expense	35.5	40.7	22.6	17.8	18.1	33.4
Off-farm income as % of total income	21.2	22.0	19.0	18.0	18.5	21.5

Source: United States Department of Agriculture, ERS, *Economic Indicators of the Farm Sector: National Financial Summary,* 1985 and 1986.

farm income for the country's farmers as a whole increased by $16,573 million in 1986 over what it was in 1974, an increase of 59 percent in just 12 years. This increase represented an average of $20,212 per farm in 1986 (Table 4.1). Off-farm income as a percent of total income decreased between 1974 and 1984 but increased in 1986. However, the increase in 1986 was still below that of 1974.

An evaluation of the financial conditions of the nation's small-scale farmers as a group shows an increase in their gross farm income between 1964 and 1982 and a decrease since 1982 (Table 4.2). The decrease between 1982 and 1986 amounted to an average of $800 per farm. On the other hand, with the exception of 1986, their production expenses increased since 1964. In fact, the increase in their production expenses between 1982 and 1984 was higher than their gross farm income. Consequently, their net farm income declined since 1964, reaching negative values in 1982 and 1984. For example, their gross farm income in 1984 averaged $15,748 per farm, while their production expenses averaged $17,067 per farm, resulting in a negative average net farm income of $1,195 per farm.

The problem of small-scale farmers' income can also be seen from the net farm income as a percent of expenses figure. Net farm income in 1964 represented about 50 percent of expenses; however, it had declined since then reaching, -4.3 and -7.7 percent in 1982 and 1984, respectively (Table 4.2). Small-scale farmers also show a declining net farm income when measured by acres of land they farm, by age of operators and by type of commodities they produce. The above result indicates that net farm income problems are more prevalent for the small-scale producers, compared to all other farms in the nation.

In order to supplement their negative farm income, most of the small-scale farmers are engaged and have become increasingly dependent on off-farm employment. For example small-scale farm operators as a whole averaged $22,534 in off-farm income in 1986, an increase of 574 percent over what it was in 1964 and 31 percent over 1982 (Table 4.2). Off-farm income in 1964 was about 29 percent of total income for that

TABLE 4.2: Farm Income and Expenses of Small-Scale Farms, 1964-1986

	Year					
Item	1964	1974	1978	1982	1984	1986
Gross farm income						
($ mil)	26,695	26,361	26,473	27,414	26,356	24,412
Average/farm ($)	8,063	11,392	14,104	15,941	15,748	15,140
Farm production						
expenses ($ mil)	17,829	21,447	25,403	28,653	28,565	24,189
Average/farm ($)	5,385	9,268	13,534	16,662	17,067	15,001
Net farm income						
($ mil)	8,866	4,914	1,070	-1,239	-2,209	223
Average/farm ($)	2,678	2,122	570	- 721	-1,195	139
Off-farm income						
($ mil)	11,067	24,983	25,118	29,660	31,627	36,336
Average/farm ($)	3,343	10,796	13,382	17,247	18,897	22,534
Net farm income as						
% of expense	49.7	22.9	4.2	-4.3	-7.7	0.9
Off-farm income as						
% of total income	29.3	48.7	48.7	52.0	54.5	60.0

Source: Ibid.

year, but by 1986, it represented 60 percent of total income. This result confirms the fact that off-farm income has been, and continues to be, a factor which affects the well-being of small-scale farm operators, particularly those with deteriorating cash flow positions and declining farm equity. Despite relatively high average off-farm income, however, nearly one-third of small-scale farm operators were below the official poverty line[3] in 1986.

Further evaluation of the nation's small-scale farm business earnings and expenses based on sales class within the group, reveals the financial conditions of the various types of small-scale farmers in the country. Of the small-scale farmers, those with sales of less than $5,000 averaged a negative net farm income since 1974, while their average production expenses per farm consistently increased for the same time period (Table 4.3). For those in the sales category between $5,000 and $9,999, and $10,000 to $19,999, the average net farm income reported was $3,250 and $5,443, respectively, in 1964. For the same time period, however, their production expenses averaged $5,923 and $11,432 per farm. This group of farmers, however, had averaged a negative net farm income since 1982, but farmers in the sales class between $20,000 and $39,999, with only one exception, had a positive net farm income since 1964. The exception was 1984 where this group of farmers had an average net farm income of -$952 per farm.

For the small-scale farm operators, off-farm income is more highly concentrated in those farms with sales less than $10,000. These farms averaged $3,400 in 1964 and $22,600 off-farm income in 1986 (Table 4.3). Although the average net farm income of these farmers was negative in 1986, their total average income, including off-farm income, was near the average for all small-scale farms. Off-farm income has also supported the farming operation of farms with sales between $10,000 and $19,999 and $20,000 to $39,999. Their off-farm income per farm, respectively, reached $19,360 and $14,252 in 1986, representing more than the value of their net farm income for the same time period.

TABLE 4.3: Average per Farm Income and Production Expenses of Small-Scale Farms by Sales Class, 1964-1986

Sales Class	Gross Farm Income	Production Expenses	Net Farm Income	Off-Farm Income
		Dollars		
Less than $ 5,000:				
1964	2,615	1,507	1,109	3,923
1974	3,906	4,147	- 238	12,756
1978	6,038	6,714	- 676	14,961
1982	8,536	9,419	- 883	18,725
1984	8,485	9,809	-1,324	20,418
1986	7,851	8,879	-1,028	26,704
$ 5,000 - $ 9,999:				
1964	9,173	5,923	3,250	2,831
1974	10,478	8,883	1,515	10,407
1978	12,188	11,774	414	13,793
1982	14,202	14,850	- 647	18,183
1984	14,580	15,617	-1,037	19,776
1986	3,961	14,061	- 100	18,400
$ 10,000 - $ 19,999:				
1964	16,875	11,432	5,443	2,058
1974	18,514	14,207	4,477	8,064
1978	20,319	18,880	1,439	12,078
1982	20,924	22,279	-1,356	16,105
1984	21,493	23,455	-1,962	17,410
1986	21,161	20,873	288	19,360
$ 20,000 - $ 39,999:				
1964	31,204	22,769	8,432	2,287
1974	35,403	25,379	9,888	6,000
1978	36,773	32,761	4,012	8,998
1982	36,966	36,577	389	12,542
1984	37,764	38,716	- 952	13,853
1986	39,099	34,206	4,894	14,252

Source: Ibid.

Trends in Farm Assets and Debts

The overall financial pictures identified above can also be shown from the balance sheet[4] of the small-scale farmers. Due to higher land values, total assets of small-scale farmers increased during the 1970s and the early 1980s. Since 1982, an economic recession and a reduced export market dropped the inflation level and net farm income drastically. As a consequence, land prices have decreased since this time with the result being a decline in total farm asset value and net worth (equity).

The value of total assets of small-scale farmers decreased by 21 percent in 1986 over what it was in 1978 and by 29 percent over 1982 (Table 4.4). Real estate, the major portion of total asset value, represented about $206,485 million of the nation's small-scale farmers' assets in 1978, $226,345 in 1982 and $152,561 in 1986. The value for 1986 was 26 percent below the amount of 1978 and 33 percent of 1982, marking the fourth consecutive period that real estate values of small-scale farmers have dropped. The major causes for the value to drop to such a low level in 1986 has been the existence of low farm income and high interest rates since 1982.

The same trend since 1982 has been observed for the non-real estate asset values. While real estate and non-real estate asset values declined between 1982 and 1986, financial assets increased by 19 percent for the same time period. Small-scale farmers' financial assets value which was $5,112 million in 1978, reached $8,699 million in 1986, accounting for about four percent of the total asset value for that year.

On the other hand, between 1978 and 1982, the total farm debt of small-scale farmers increased by almost six percent. Of this total farm debt, over one-half was in real estate (Table 4.4). Non-real estate debt increased by 20 percent between 1978 and 1982. Although there has been an increase in total farm debt between 1978 and 1982, the overall debt of small-scale farmers in the country has declined since 1982. This is because of a slowdown in the expansion of real estate debt due

TABLE 4.4: Farm Balance Sheet of All Small-Scale Farmers, 1978-1986[a]

Item	Year				
	1978	1980	1982	1984	1986
Total assets ($ mil):	256,667	283,777	288,016	249,127	203,374
Real estate	206,485	226,311	226,345	191,979	152,561
Non-real estate	45,070	53,162	54,373	49,642	42,114
Financial	5,112	4,304	7,298	7,506	8,699
Total debts ($ mil):	38,057	38,804	40,144	38,660	31,193
Real estate	23,693	24,700	22,889	23,150	19,496
Non-real estate[b]	14,364	14,104	17,255	15,516	11,697
Total equity ($ mil):	218,612	244,974	247,873	210,464	172,183
Total debt-to-asset ratio	14.8	13.7	14.0	15.4	15.3
Total debt-to-asset ratio (U.S.)	14.6	13.5	19.3	21.0	20.7

[a]Excluding farm households.
[b]Including Commodity Credit Loans.

Source: *Economic Indicators of the Farm Sector: National Financial Summary,* 1981, 1985 and 1986, ERS/USDA, Washington, D.C.

to high real interest rates, low returns on capital invested in agriculture and lower land values.

Since debt of small-scale farmers did not decline as much as their asset values, the effect has been an increase in debt to asset ratio.[5] Table 4.4 shows the debt/asset ratio of these farmers between 1978 and 1986. This ratio, which is one of the most widely used measures of financial strength or weakness, has increased consistently since 1982 reaching 15.3 percent in 1986, representing the second highest level of increase since 1978. This means that relative to the value of total farm assets, claims in the 1980s have increased for the small-scale farmers. However, the increase in this ratio has been below that of the national average (Table 4.4).

The relative increase in debt and decrease in asset values also resulted in equity decline of small-scale farmers in the nation. In 1986, their total farm net worth fell to about $172,183 million, the lowest level of owner equity since 1964 (Table 4.4). This drop again was largely due to the drop in land values, because real estate accounts for nearly 75 percent of total farm asset values. This is also the major reason for the increase in the debt to asset ratio in each year since 1980. However, owner equity in the coming years in the country is expected to increase. The expectation is based on the relative stability trend observed in both real estate and non-real estate asset values and further reduction in the level of debt outstanding.

The nature of small-scale farm operators' assets and debts varies by size of farms within the group. Among the small-scale farmers, those with sales of less than $5,000 had an average real estate value of $80,586 and non-real estate value of $16,911 per farm in 1982 (Table 4.5). However, by 1986 it declined to $60,144 and $14,855, respectively. The trend was similar for the other categories of small-scale farmers. For example, in 1982 farmers with sales between $10,000 and $19,999 averaged real estate value of $174,144 and those between $20,000 and $39,999 averaged $266,141. But by 1986 it declined to $130,038 and $198,801 per farm, respectively. The non-real estate asset values of these categories have also declined between 1982 and 1986. They fell 13 percent for

TABLE 4.5: Average per Farm Balance Sheet of Small-Scale Farmers by Sales Class, 1982-1986

Sales Class	Assets: Real Estate	Assets: Non-Real Estate	Debts: Real Estate	Debts: Non-Real Estate	Total Equity	Debt/ Asset
	Dollars					(%)
1982:						
Less than $5,000	80,586	16,911	8,059	4,530	87,991	12.9
$5,000 - $9,999	121,266	29,808	11,437	8,380	135,373	13.1
$10,000 - $19,999	174,144	43,989	17,791	14,385	191,021	14.7
$20,000 - $39,999	266,141	68,836	27,957	25,398	289,177	15.9
1986:						
Less than $5,000	60,144	14,855	7,616	3,725	67,836	15.1
$5,000 - $9,999	90,514	26,344	10,800	6,616	104,494	10.4
$10,000 - $19,999	130,038	38,193	16,799	10,698	147,073	16.3
$20,000 - $39,999	198,801	57,769	26,399	18,374	221,431	17.4

Source: Ibid.

those between the sales class of $10,000 and $19,999, and 16 percent for those between $20,000 and $39,999.

In contrast, total outstanding farm debt per farm for each category of small-scale farmers decreased between 1982 and 1986 (Table 4.5). However, both in 1982 and 1986, real estate debt was higher than non-real estate debt. For farms with sales less than $5,000, farm real estate debt in 1986 averaged $7,616 per farm, a decrease of five percent since 1982. Farm real estate debt outstanding of those between $5,000 and $9,999 averaged $11,437 per farm in 1982, while the average for 1986 was $10,800.

The decreasing trend in farm real estate debts between 1982 and 1986 was also true for those in the sales range between $10,000 and $19,999 and $20,000 to $39,999. The decrease in farm real estate debt observed in Table 4.5, was due to debt repayments as well as loan losses that occurred between 1982 and 1986. In addition to farmers' desire to pay off debt, the decrease in outstanding real estate debt came from lenders' reluctance to further extend new debt and the charge-off of losses as lenders wrote off uncollectible loans.

Non-real estate debt for the various sales categories of small-scale farmers also dropped since 1982, accounting for an average of about 18 percent per farm for the sales class less than $5,000, 21 percent for $5,000 to $9,999, 26 percent for $10,000 to $19,999, and 28 percent for the sales class between $20,000 and $39,999. The decrease in the non-real estate debt during these periods is due to various reasons among which are delayed capital investments, reduced planted acreage, and lower production expenditures. In addition, requirements of real estate as collateral for short-term loans by some lenders have contributed to the decline of non-real estate debt.

On the other hand, the net worth (total equity) of each small-scale operator has decreased since 1982 (Table 4.5). For example, on the average, farms with sales between $10,000 and $19,999 lost $43,948 per farm between 1982 and 1986, while the loss amounted to $67,746 per farm for farms with sales between $20,000 and $39,999. The trend was similar for the sales categories of less than $5,000 and $5,000 to $9,999. Most of the decline in equity can be attributed to the continuing erosion of farm real estate asset values observed since 1982. On the average, however, farms between $10,000 and $19,999 and $20,000 to $39,999 sales categories have more equity than those with sales less than $5,000 and those between $10,000 and $19,999. This strong positive relationship between net worth and size of farm is historically consistent for the farm sector of the U.S.

The average debt/asset ratio for all small-scale farm operators, as observed in Table 4.4, increased between 1982 and 1986. With the exception of farms in the sales category of $5,000 to $9,999, the increase in the ratio has been true for all other sales class categories of small-scale farmers. For example, for those farms with sales less than $5,000, the ratio was 15.1 percent in 1986 compared to about 13 percent in 1982. For those with sales between $10,000 and $19,999 the ratio increased from 15 percent in 1982 to 16 percent in 1986.

Among all small-scale farm operators in the country, those between $20,000 and $39,999 have a slightly higher percentage debt both in 1982 and 1986. Overall, however, farms with sales less than $40,000 had both a lower proportion of farms in the worst relative financial position and more in a favorable position in 1986 than farms with sales over $40,000 (Table 4.5).

Further insights into the financial performance of farm operators by sales class for the nation is presented in Table 4.6. The table shows selected financial (debt/asset) ratios[6] to help us identify the types of small-scale farmers that are in a vulnerable position and those that are not. As can be seen from the table, the large sales class has the highest proportion of farm with debt/asset ratios in excess of 40 percent.

For example, 41 percent of farms with sales over $500,000 have debt/asset ratios over 40 percent, and about 39 percent of the farms with sales between $250,000 and $499,999 have debt/asset ratios over 40 percent. In contrast, about 20 percent of the farms with sales between $20,000 and $39,999 have debt/asset ratios over 40 percent. The percentage distribution for those sales class from $10,000 to $19,999 and for those with less than $10,000 in sales was 16 percent and 12 percent, respectively (Table 4.6).

TABLE 4.6: Distribution of Farms by Debt/Asset Ratio and Sales Class, 1986

Sales Class	No Debt	Debt/Asset (1% - 40%)	Debt/Asset (40% - 70%)	Debt/Asset (over 70%)
		Percent of Farms		
Less than $10,000	55.7	32.4	7.8	4.0
$ 10,000 - $ 19,999	48.6	35.2	9.5	6.6
$ 20,000 - $ 39,999	35.7	44.4	13.3	6.6
$ 40,000 - $ 99,999	21.9	45.4	17.7	14.9
$100,000 - $249,999	16.2	48.2	20.5	15.1
$250,000 - $499,999	10.8	50.4	23.3	15.5
$500,000 and over	13.2	46.1	24.5	16.1
All Farms	38.9	39.4	13.0	8.6

Source: *Financial Characteristics of U.S. Farms, January 1, 1987,* ERS/USDA, Washington D.C.

Historically, a debt/asset ratio of 70 percent or greater was cause for concern for the farmers' well being. Today a value greater than 40 percent creates financial problems for most farmers. The data in Table 4.6, therefore, suggest in general that given the levels of interest rates and net returns in farming, farms with debt/asset ratios of over 40 percent will have some difficulty repaying their debt. It also suggests that farms with debt/asset ratio over 70 percent are almost certain to have serious cash short-falls.

Farms with debt greater than their assets (debt/asset ratios over 100 percent) are generally considered to have little, if any, chance of survival. For the small-scale farmers as a whole, farmers in sales category of $10,000 to $19,999 and $20,000 to $39,999, each accounted for about 7 percent of farmers having debt to asset ratios in excess of 70 percent. Of these farmers about 3 percent represented farmers with debt/asset ratio of 100 percent. On the other hand, those less than $10,000 had the lowest (4 percent) debt to asset ratio in excess of 70 percent in 1986. However, about 50 percent (2 percent of this 4 percent) represented farmers with their debt greater than their assets, i.e., debt/asset ratio over 100 percent.

Operator Financial Stress

The above comparisons suggest the likely trends of the financial status of all farms, particularly small-scale farms in the U.S. That is, the greatest financial stress likely occurs among those farms with annual sales of over $40,000. Farms with annual sales less than $40,000, representing about 72 percent of the farms in the country, have the smallest debt loads.

However, using the net farm income data in Table 4.3 with the debt/asset ratio data in Table 4.6, and based on the following criteria, it is possible to identify the types of small-scale farmers who are under financial stress or in a favorable position. Farms are identified to be in a favorable position if they have positive net farm income and low debt/asset ratio (less than 40 percent). If they have negative net farm income and high debt/asset ratio (over 40 percent), they are identified to be in a vulnerable position. On the other hand, farms with low debt/asset ratio but negative net farm income are classified to have marginal income (not generating sufficient income), while those with high debt/asset ratio and positive net farm income are classified to be marginally solvent (not experiencing short-term income difficulties).

Based on the above criteria Table 4.7 was constructed to show the distribution by income-solvency position of small-scale farmers in 1986. As shown in the table, about 9 percent of all small-scale farmers were financially vulnerable. On the other hand, about 60 percent of these farmers were in a favorable financial position, while 24 percent had a secure debt/asset ratio, but with net farm income that was not sufficient to cover their expenses.

Among the small-scale farmers in 1986, the sales class with the largest percentage of farms in a vulnerable and insolvent financial position were those between $10,000 and $19,999. In contrast, farms with sales less than $10,000 appeared to be in a less vulnerable financial position, while they represented the highest percentage of those small-scale farmers with problems of generating cash flow (marginal income position). The small-scale farmers in the sales class between $20,000 and $39,999, on the other hand, are found to be in a marginal solvency position.

These farmers are in a marginal solvency position because they were able to generate cash flow to stay in business even though they had high debt/asset ratio. Assuming costs and revenues remain the same, their ability to generate cash flow will likely keep these highly leveraged farmers in business for the coming years. However, if costs were to rise in relation to revenues or if returns were to drop due to changes in production, prices, or other reasons, this class of farms could move into a more difficult financial situation, hence increasing the number of small-scale farmers in a vulnerable position.

As indicated above, small-scale farmers as a group have less financial stress than those with sales above $40,000. Among other things, the following are partial explanations for this situation: (1) operators of these farms pay their loans primarily from their off-farm income rather than income from farming; (2) small-scale farmers lack the interest to borrow

TABLE 4.7: Distribution of Small-Scale Farm Operators by Net Farm Income and Solvency Position, 1986

	Income/solvency position				
Sales class	Favorable	Marginal Income	Marginal Solvency	Vulnerable	Total
			Percent		
Less than $10,000	61.2	27.0	4.5	7.4	100
$10,000 - $19,999	59.0	24.8	6.2	10.0	100
$20,000 - $39,999	61.1	19.1	10.1	9.7	100
All Farms	60.5	23.6	6.9	9.0	100

Source: Ibid.

capital, for expansion or production purposes, as they wish to remain debt-free because of risk considerations; and (3) those who perceive credit financing as an essential factor in farming continue to have problems getting farm credit from lending institutions.

Farm Debt by Lender

Most small-scale farmers traditionally have financed the major share of capital requirements for farming operations from internal savings (equity capital). Although there is a history of a low borrowing rate by small-scale farmers, they have overwhelmingly characterized and perceived credit financing as an essential functions in farm business (1). In recent years, those who saw credit as an essential factor to expand their business and to survive have often been disqualified from farm credit loans by various lending agencies because of their disadvantaged economic conditions.

In the U.S., the major providers (holders) of farm credits (debts) are the commercial banks, the farm credit system, Farmers Home Administration, and individuals and others. American farmers obtained about 21 percent of their total agricultural credit from individuals and others (Table 4.8). For example, in 1986, individuals were the third largest lenders of non-real estate (production loans) debt. Therefore, any financial problems faced by farmers, large or small, in the country are major concerns for individual lenders. Also, this group of lenders along with the farm credit system were the primary source of real estate loans, accounting for a combined total of 37 percent of the real estate loans in 1986. These institutions, therefore, were most severely affected by declining land values.

Lending from the Federal Land Banks represented about 23 percent of the total farm real estate loans in 1986, an increase over what it was in previous years. Borrowing from the Federal Land Banks has been substantially increasing within the last few years for two reasons. First, the Federal Land Banks have the lowest cost funds because of the Federal

TABLE 4.8: Distribution of All Farm Debt Outstanding by Lender in the U.S., 1986

Lender	Type of Debt		
	Real Estate	Non-Real Estate	Total
		Percent	
Commercial Banks	7.6	18.7	26.3
Farm Credit System:	22.6	6.6	29.2
Federal Land Banks	22.6	n/a	22.6
Production Credit Associations	n/a[a]	6.4	6.4
Federal Intermediate Credit Banks	n/a	0.2	0.2
Farmers Home Administration	6.2	9.8	16.0
Life Insurance Companies	6.6	n/a	6.6
Individuals and Others	14.4	7.4	21.8
Commodity Credit Corporation	0.1	n/a	0.1
Total	57.5	42.5	100

[a]Not applicable.

Source: *Economic Indicators of the Farm Sector: National Financial Summary*, 1986, USDA/ERS.

Credit Administration (FCA) policy of basing interest rates on their average cost of funds. Commercial bank interest rates are much higher than any of the institutions in the federal credit system. Second, the restructuring of short-term debt into long-term debt begun in 1980 to obtain more favorable repayment terms continued until 1986 (2).

The Farmers Home Administration held 16 percent of all outstanding farm debt in the country in 1986, with its non-real estate debt being slightly higher than real estate debt. Commercial banks lending secured by farm real estate accounted for only eight percent, while 19 percent of the non-real estate loans were provided by these banks, making them the dominant lenders of non-real estate loan for the year. Real estate farm debt of the Commodity Credit Corporation accounted for only 0.1 percent, representing the smallest percentage in the real estate loans outstanding of all lenders in 1986. On the other hand, at 0.2 percent the Federal Intermediate Credit Banks accounted for the smallest non-real estate loans for the same time period. Overall, about 58 percent of the loans by various lenders in the U.S., in 1986, was made for real estate purposes while 42 percent was for non-real estate purposes.

The distribution of farm debt also varied by sales class of farms. For examples, for farms with sales less than $10,000 and between $10,000 and $19,999 a large share of their debt (loan) was provided by commercial banks and less from the farm credit system. In contrast, farms in the sales class of $20,000 to $39,999 and $40,000 to $500,000 had their largest portion of loans provided by the farm credit system in 1986. Farm loans for the largest sales class (over $500,000) were nearly equally divided between the farm credit system, commercial banks and other sources of funds, with the latter being slightly higher.

All lenders in 1986 reported to have at least 59 percent of their debt held by highly leveraged farm operators. Of these the Farmers Home Administration had the highest percentage, while individuals and others had the lowest share. Although all farm lenders have been affected by the farm financial difficulties of recent years, as indicated above, some have been more vulnerable than others. For example, the farm credit system differs from other agricultural banking systems in several respects. The farm credit system only serves the agricultural sector, making it more vulnerable than most agricultural banks to downturns in the nation's farm economy. In 1986, 23 percent (larger than any other bank loans) of the outstanding loans of the farm credit system were real estate loans. Although small in numbers, these financial institutions suffered large losses due to the fall of land values.

Another factor that makes the farm credit system different from other agricultural banks is the nature of its national charter. The charter calls for the system to provide an on-demand lending service to eligible and credit worthy applicants in all geographic areas under all economic circumstances, while other banks have more flexibility in determining their lending business. The above are only two examples of the differences of the farm credit system relative to other credit banks serving agriculture, to illustrate the importance of understanding the current situation of the lending institutions in the country.

Notes

1. Net farm income is the difference between gross farm income and total expenses, and measures the net value of agricultural production for a given year.

2. USDA defines off-farm income as income received by farm operators and their households from nonfarm wages and salary jobs, wages, and salaried earned on other farms, nonfarm businesses and professional income, interest and dividends, and all other cash nonfarm income.

3. The poverty line is an estimate of the minimum income level necessary to cover essential living expenses based on size of family. In 1986, the poverty threshold for a family of four was $11,200.

4. The farm sector balance sheet measures or estimates current market values of total assets, debts (liabilities) and owner's equity as of December 31 of a calendar year.

5. Debt/asset ratio is generated by dividing the total debt (liabilities) of a farm business by its total assets, and it indicates or measures the long-run financial strength (weakness), or relative indebtedness of the farm business. The lower the ratio, the better the financial position of the farm business since a smaller proportion of assets are owed to creditors.

6. Farmers are defined to be without serious financial difficulty if the ratio is less than 40 percent. Between 40-70 percent, they are feeling moderate financial problems, and over 70 percent, they are technically insolvent (i.e., highly leveraged).

References

1. Huffman, D. C., et al. "Socio-Economic Characteristic and Income Opportunities of Small Farms in Selected Areas of Louisiana," *Agricultural Economics Research Report, No. 580,* DAE/LSV, Baton Rouge, Louisiana, 1981.
2. U. S. Department of Agriculture. *Economic Indicators of the Farm Sector: Farm Sector Review,* 1986. ERS/USDA, Washington, D.C.

5

Factors Affecting the Growth and Development of Small-Scale Agriculture and Rural Communities

The trend toward fewer but larger farms that has been observed in Chapter 1, the analysis of small-scale farm characteristics in Chapters 2 and 3, and their financial conditions described in Chapter 4, are the result of interactions and changes of various economic and non-economic causal factors or constraints. Some of these constraints are resource endowments, education, appropriate technologies, marketing, and government programs and policies.

The interaction and changes of these factors have affected all sizes of farms, but the effect has been greater on small-scale farms, particularly on black-owned and operated farms. The following are explanations of how these factors have shaped and continue to shape the growth and development as well as the survival of small-scale farm operators in the U.S.

Resource Endowment

The findings of this book have shown that small-scale farm operators averaged smaller farm sizes and less farm income than large-scale farm operators. The size and income of black operated farms in particular, has been, and continues to be, smaller than those operated by white farmers. Therefore, over the years the ownership and control of land and other resources, such as capital, were concentrated in the hands of large-scale and white farmers.

Small-scale farmers started with less land; as a result, they had to work a larger percentage of their land each year. This intensive use of their farmland caused the soil to become depleted and less productive. The result of low productivity has been low farm income. Under conditions of low income and low productivity, the pressure to maintain a basic subsistence living deterred many small-scale farmers from making substantial long-term investments. Therefore, small-scale farmers were unable to adopt capital intensive production practices.

The major share of the capital requirements of small-scale farming operations has come from internal financing (equity capital). However, their equity capital (which is strongly and positively related to size, Chapter 4), has never been sufficient to allow these farm operators, particularly black operators, to maintain a competitive position in farming. To survive, expand operation, and maintain a competitive position, a farmer must be able to borrow capital to purchase production inputs such as seeds, fertilizer, machinery and equipments, just to mention a few.

However, many small-scale producers were often constrained by their ability to secure credit. Furthermore, those operators who saw credit financing as an essential factor in maintaining a competitive position as well as expanding their operation were rejected by lending institutions. This was done at the time when no shortages of loanable funds in the farm sector existed or were evident. Even when small-scale producers were able to borrow, they were charged high interest rates since they were considered to be high-risk borrowers.

According to Horne (1) many lending institutions seek only larger borrowers, in order to minimize their service costs per dollar loaned. To obtain a loan, the small-scale producer may have to pay a higher rate of interest. Since most small-scale farmers possess limited information about available sources of credit, they usually do not compare interest charges or other measures of credit's true cost. The terms of loans to small-scale farmers are likely to be tailored to the lender's convenience and profitability instead of being mutually beneficial.

Lenders may help the large operator prepare a cash flow and a projected profit and loss statement. Since the small-scale farmer is not a preferred customer, he probably will not receive that service. Consequently, small-scale farmers often face the disappointment of being rejected for a major loan because of insufficient attention to planning. A small-scale farmer whose knowledge of sources of credit is limited may become so frustrated he will never try to borrow again, thereby cutting himself off from the potential of higher income based on wise borrowing (1).

The credit situation was worse for black farm operators. In addition to charging high interest rates, some of the lending institutions in the past insisted on the type of crops to be produced (tobacco and cotton) by black farmers. However, when markets for these products failed and/or production practices changed, as was the case in the 1920s and 1930s, most of the black farm operators were unable to pay their credit debts (2). Faced with no other alternatives and an impossible debt payment, many of these farmers sold their land and left agriculture. For those who decided to stay in farming, credit problems have continued to plague them until today.

For most small-scale farmers, particularly blacks, the only resource endowment that was adequate, if not in surplus, was labor. Although this group of farmers possessed adequate labor, they lacked the management skills or the training necessary to operate the farm business successfully. The success or failure of a farm business, however, is dependent upon the operator's ability to manipulate the farm's resource mix (land, labor and capital). In other words, the optimal combination of land, labor, capital and management is necessary to achieve maximum economic return.

Education

Education has a direct impact on the management ability of a farmer. A study by Tweeten suggested education to have a high economic payoff to individuals in most areas and occupations (3). He further stated that "it is highly profitable to individuals in rural poverty areas who have geographic and occupational mobility, but it is likely to be only marginally profitable to those lacking mobility" (3). Although limited capital and lack of economic opportunities in low income rural areas limit the productivity of education for persons who remain in the rural area, Tweeten indicated that quality education still offers the only real hope for increasing income.

However, the quality and quantity of education received by most small-scale farmers in the past, particularly blacks, was inadequate to bring about greater equality in economic opportunities and income. Among other things, the lack of quality education made most of the small-scale farm operators less able to: (1) understand, and hence, utilize or adopt, available technologies; (2) manage the financial affairs of their business; and (3) acquire and interpret farm related information (4,5,6).

A recent report of the Southern Growth Policies Board Commission listed ten regional objectives to be achieved by 1992 in the South, where most of the small-scale farms are located. Of these, seven were directly related to education: (1) provide a nationally competitive education for all Southern students; (2) mobilize resources to eliminate adult functional literacy; (3) prepare a flexible and globally competitive workplace; (4) increase the economic development role of higher education; (5) increase the South's capacity to generate and use technology; (6) develop pragmatic leaders with global vision; and (7) improve the structure and performance of state and local governments (7). Education, therefore, is an extremely important factor for the survival and human resources development of small-scale producers.

Almost 95 percent of the nation's black farm operators live in the South. The educational opportunities for blacks in this region have been marked by segregated academics and poorly-financed public school systems. Because of racial barriers, black schools were not funded at the level of whites; black teachers were not paid as well as whites; and the school year for blacks was not as long as for whites (8). For example, some of the traditional black Land-Grant institutions were in existence as early as 1862 and the rest since 1980. However, Congress did not appropriate any federal funds for these institutions until 1972. In contrast, the white Land-Grant institutions, founded in 1862, have been receiving federal funds since 1887 (9).

All these factors, therefore, worked together to prevent blacks from receiving quality education as well as weakening their competitive position in farm business. As a result, black farm operators left agriculture at a much faster rate than whites. If the objective of the federal government is to stop or at least to slow down the rate of decline, as well as to improve the well-being of this group of farm operators, special attention must be given to their educational needs.

Technology

In the history of U.S. agriculture, a lot has been accomplished in the development of new agricultural technologies. Unfortunately, many small-scale farmers were unable to adopt many of these technologies. Lack of knowledge, limited resources, limited managerial as well as risk bearing ability are some of the reasons that have affected the adoption of technologies on all small-scale farms. By and large, the early adopters, hence, the beneficiaries of these technologies, were the operators of large farms.

As explained in the previous chapters, small-scale farm operators in general, and black farmers in particular, control limited resources of land, capital and management skills. Because of these resource constraints, small-scale farmers often have been unable to adopt new technologies. Even when they do, it has been more slowly (late adopters) than large-scale farm operators in the country. For those small-scale operators who, although slow, elected to adopt some of the new technologies (such as machinery and equipment), the result has been an over investment, which led to an increase in their cost of production.

These farmers could have taken the alternative approach of buying used machinery and equipment. However, this alternative possibly would bring with it higher repair costs, as well as more down time. In addition, by the time they decided to adopt these innovations, they often find that output among all farmers has increased. As a result, without an accompanying increase in aggregate demand, they found prices for their products declining.

Rate of adoption of new technologies is also influenced by the availability of labor and education. For the most part, the technologies in agriculture have focused on reducing labor requirements. Small-scale farm operators with adequate labor and with skills not readily marketed in off-farm employment were usually slow to adopt these new labor-saving technologies. Today, as in the past, the key element in increasing productivity is substitution of capital for labor. However, due to limited capital, surplus labor and prohibitively high input prices, small-scale operators are forced to use fewer inputs and consequently have lower yield and poorer quality products (10).

Smallness in and by itself has been a disadvantage in the marketplace of input for small-scale producers. For example, a study by the U.S. Chamber of Commerce, one of the strongest supporters of large-scale producers, found that large-scale farms have been able to buy many of their inputs (except labor and management) for 15 to 25 percent less than the price paid by more small-scale farmers (11). Discounts were not justified to be given to small order buyers of inputs by agribusiness firms.

Furthermore, not only is the price of inputs prohibitive for small-scale farmers, but in some cases these inputs (machinery, equipment) were not designed or sized according to the scale or labor situations of these farms. In order to increase income on small-scale agriculture and reduce their cost of production, there is a need for new and innovative technologies that are sized to meet the needs of these farms, affordable, easy to maintain, and free of unnecessary items or gadgets.

Marketing

Another factor that accounts for the general inability of small-scale farmers to maintain a viable operation is the lack of efficient[1] output marketing alternatives. Marketing is extremely important for any farm product. In fact, it is often said that marketing is the single most important element in a farmer's entire business. However, marketing agricultural products has been a problem that has persisted throughout most of American agricultural history. Today, it continues to be a problem for all agricultural producers, particularly for the operators of small-scale farms.

Marketing is a particular problem for these farms because of the structure of markets developed in response to the new technologies which restricted marketing access for this group of farmers. For example, technological developments on the farm, in transportation, refrigeration, storage, packaging and communication contributed to a concentrated marketing system. These developments brought a mass retailing system which, in turn, required a large volume of standardized products to be available. These requirements thus significantly eliminated the small-scale farmer from the mass retailing market system.

The majority of small-scale farmers produce traditional products (tobacco in the case of black farmers) within the local area; as a result, they are limited in marketing alternatives and bargaining strength due to low volume (12). Even production of non-traditional products (such as exotic crops) poses similar problems. Again, as small volume producers they alone can not attract a market.

Marketing agricultural products is also a problem to small-scale farmers due to their lack of education. This restricts their ability to receive and utilize marketing information, which is a is a vital force in making a sound management decision. Furthermore, given the size of these farms, the available marketing information, in many cases, is not well suited to the needs of small-scale farms.

Small-scale farms are further disadvantaged in marketing their products due to lack of storage facilities. For example, small-scale producers may receive, like the large-scale farmers, high prices for grain if they have the facilities to store the product on their farms beyond the period immediately after harvest, when prices are usually low, until a time when prices are higher.

As a result of these problems, small-scale farmers have been forced to seek alternative means of marketing their products to gain access to market outlets for their products. In recent years, the interest in marketing by small-scale producers has centered around the direct farm-to-consumer marketing alternatives. These markets are identified as road-side, farmers', pick-your-own, and pooling or cooperative marketing operations. These marketing alternatives have particularly become increasingly viable in the South as industry moves into rural areas and population density increases. The development of these kinds of direct marketing takes time and effort; however, it has proven to be profitable for some small-scale farmers (13).

Management and Information

The survival of most small-scale agricultural operators is dependent on the farm operator's ability to generate from the farm enterprise adequate annual income, which is often the main source of family income. The ability to generate income is dependent upon the management skills of the operator to manipulate the farm's resource mix (land, labor and capital). However, small-scale producers lack the management skill or the training necessary to operate their farm business successfully and achieve maximum economic return. In fact, many researchers concerned with problems of small-scale producers agree that the failure to effectively use efficient management practices is one of the primary factors limiting improvements in their farming operations (14,15,16,17).

Another important factor associated with failure to use effective management practices by small-scale farmers is the lack of adequate information. Adequate information, however, is important to enable a farm manager or operator to identify niche market opportunities, produce a product that meets consumers' specification and satisfaction, and respond to consumer feedback change in the product.

Today we are making the transition into yet another era which can conveniently be labeled the information technology age. This information technology age is based upon dramatic improvements in computers and telecommunications, the marriage of which is producing profound improvements in a manager's ability to identify, sort, retrieve, transmit, create, and apply useful information. As in the past, small-scale farm operators of today lack adequate sources of information to fully participate in the existing market structure.

Government Policies and Programs

Although farm programs were designed to benefit all farm sizes, and particularly small-scale farms, they have disproportionately benefited large-scale farms because most government commodity programs and policy efforts to stimulate production in the past were directed to the large commercial farmers (18). With efficient use of economic principles, adoption of new technology, and substitution of capital for labor, the large-scale farmers dominated the production and market shares over the years. There is little doubt, therefore, that government policies and programs have affected the structure of agriculture and have led to fewer but larger farms (by displacing small farmers) in the U.S. (18,19,20).

Major components of government policies and programs used by the federal government were supply control through restrictions; marketing quotas; price support through direct purchases of commodities; use of non-recourse price support loans; acreage allotments; and long-term land retirement (soil bank program). For example, under the direct cash payments program, farmers were paid directly to reduce the acreage planted in certain crops. Under the acreage allotment program, farmers were restricted to a fixed acreage that can be planted in a particular crop each year to receive benefits.

Both programs were, however, circumvented by increasing production on the allowable acreage, using productive inputs. Furthermore, marginal lands with poor soil were taken out of production and the allotted crop grown only on the best land, to increase production. The overall effect of the programs was to benefit the large-scale farmers with larger allotments and more productive land. A study by Tweeten concluded that government farm programs very often redistributed income from relatively low income taxpayers to well-to-do farmers (21).

In addition, the work done by national research and extension agencies in the past, particularly the 1862 Land-Grant universities, served the interest of large-scale farmers rather than small-scale farmers. For example, federally subsidized research at the 1862 Land-Grant universities on the profitability of mechanical cotton pickers, was not done for small-scale producers but for large-scale producers, particularly when the federally subsidized cotton allotment program took effect (22).

With respect to extension, a 1972 study found that 86 percent of an extension service workers' time was devoted to helping the wealthiest third of America's farmers. The study further stated that on a per capita basis, wealthy farmers received 14 times as much attention from the extension service as low-income small-scale farmers (22). However, the main reason that the extension service was created as an agency was to disseminate the results of agricultural research to rural and small-scale farmers so that they can solve problems they face.

One of the major inadequacies of government policies and programs is that they are tied to productive resources such as land rather than to farm income need. This, therefore, means that the more control and ownership of productive farm resources a farmer has, the greater the benefits from these programs and policies he receives. Automatically large-scale farmers benefited more than small-scale and low income farmers, particularly blacks, from these farm programs. It does indeed sometimes appear that the U.S. agricultural programs and policies came from Matthew 13:12; "For whosoever hath, to him shall be given, and he shall have more abundance, but whosoever has not, from him shall be taken away even that he hath."

Given that black farmers control limited quantities of land, capital and educated human resources, they benefited slightly or not at all from government farm programs and policies of the past 40 years. A 1982 report entitled "The Decline of Black Farming In America" by the U.S. Civil Right Commission concluded that government programs, starting with the Agricultural Adjustment Act of the 1930s, did not provide blacks with benefits equal to whites. However, these programs were intended to serve low income farmers including black owner operators, tenants, sharecroppers, and farm laborers by making financial assistance available to purchase land and equipment so that they could become viable farming operations (23). Other studies have also shown that these programs had far more particularly disastrous effects on black farm operators than on whites (24,25). In fact, the nearly 96 percent reduction in the number of black farm operators since 1920 offers a mute testimony to the country's farm programs and policies insensitivity to the plight of blacks.

Sociological Constraints

The growth and development of small-scale producers could be enhanced if conditions of favorable and progressive attitudes existed. However, the gearing of technology, extension and research programs, credit systems, educational opportunities, marketing, and government programs and policies to the needs of large-scale farms have created attitudinal constraints or problems for the small-scale farmer. Since these developments and programs have met only a few of these farmers' needs in the past, even genuine assistance that is offered to small-scale farmers today is met with skepticism. In fact, today's small-scale producers are seldom actively seeking assistance. And when they do, they rely mostly on friends, family, and neighbors.

Traditional ways of doing things also play a big role in the day-to-day management of the small-scale farm. Technologies have been very slow in replacing old techniques that have been handed down for generations. And, unfortunately, many small-scale low income farmers have never travelled out of their own area enough to see other ways of doing things. In talking with them personally, one can often discover that they are very interested in using new methods, but do not believe the methods could work on their farms.

Even though their incomes from farming is low and recently negative, studies have found that most small-scale farmers would stay in farming even if they were given the opportunity to leave the farm (26,27). Some small-scale farmers see farming as the only way available to increase their income, and others do not want to be relocated since they are relatively in the advanced age category with low educational achievements. Therefore, all programs, policies and efforts should also be directed towards changing the attitudinal problems that exist today in small-scale agriculture and rural communities.

Notes

1. Many markets are "inefficient" because of the inordinate profits the "middleman" takes between producer and consumer.

References

1. Horne, J. E. *Small Farms: A Review of Characteristics, Constraints and Policy Implications.* Southern Rural Development Center, Report No. 33, 1979.
2. Marable, M. "The Land Question in Historical Perspective: The Economics of Poverty in the Blackbelt South, 1865-1920," in *The Black Rural Landowner--Endangered Species.* Ed. Leo McGee and Robert Boone, Westport, Conn.: Greenwood Press, 1979.
3. Tweeten, L. G. *The Role of Education in Alleviating Rural Poverty.* Agricultural Economics Report, No. 114, USDA, Washington, D.C., 1967.
4. Brayield, A. H., et al. "Attitudes, Interests and Personality Characteristics of Farmers," *Journal of Applied Psychology,* 41 (1987).
5. Hobbs, D. J., et al. *The Relation of Farm Operator Values and Attitudes to Their Economic Performance.* Department of Economics, Report No. 3, Iowa State University, 1964.
6. Huffman, E. C. "S Technique for Classifying Farm Managers According to Managerial Ability," *Dissertation Abstracts,* No. 24, Ohio State University, 1963.
7. Southern Growth Policies Board. *After the Factories: Changing Employment Patterns in the Rural South.* Research Triangle Park, North Carolina, 1985.
8. Huffman, W. "Black-White Human Capital Differences: Impact on Agricultural Productivity in the U.S. South," *American Economic Review,* 71 (1981).
9. Williams, T. T. "Teaching, Research and Extension at Predominantly Black Land-Grant Institutions," in *Human Resources Development in Rural American: Myth or Reality.* Ed. by T. T. Williams, Human Resources Development

Center, Tuskegee University, Tuskegee, Alabama, 1986.

10. Brown, A. *Factors Affecting the Success of Small Farms in Louisiana.* Dissertation, Louisiana State University, 1983.
11. United States Chamber of Commerce. *The Changing Structure of U.S. Agribusiness and Its Contribution to the National Economy.* Washington, D.C,, 1974.
12. Marshall, R., et al. *Status and Prospects of Small Farmers in the South.* Southern Regional Council, Inc., Atlanta, Georgia, 1979.
13. West, J. G. "Agricultural Economics Research and Extension Needs of Small-Scale Limited Resource Farmers," *American Journal of Agricultural Economics.* 11 (1979).
14. Atkinson, R. "Can Small Farmers Be Helped to Change? Yes or No?" Southern Rural Development Center, *Rural Development Research and Education,* 4 (1977).
15. Johnson, R. G., et al. *Use of Credit and Other Resources By Low-Income Farmers.* Department of Agricultural Economics and Agribusiness, Research Report, No. 472, Louisiana State University, 1974.
16. Martin, L. R., et al. "Effects of Different Levelly of Management and Capital on Incomes of Small Farmers in the South," *Journal of Farm Economics,* 42 (1960).
17. Parks, A. L. *The Socio-economic Element as a Dominant Factor in the Resource Status of the Limited Resource Farm Household.* Paper presented at the Annual Meetings of the Southern Association of Agricultural Scientist, Houston, Texas, 1978.
18. Vogeler, I. *The Myth of the Family Farm: Agribusiness Dominance of U.S. Agriculture.* Boulder, Colorado: Westview Press, 1981.
19. Bonner, J. T. "Distributional Issues in Food and Agricultural Policy," in *Increasing Understanding of Public Problems and Policies,* Farm Foundation, Oak Brook, Illinois, 1984.
20. Tweeten, L. G. *Causes and Consequences of Structural Changes in the Farming Industry.* National Planning Association, Food and Agricultural Committee, Washington, D.C., 1984.

21. Tweeten, L. G. "Commodity Programs for Agriculture," in *Agricultural Policy in Affluent Society.* Ed. by Vernon, et al., New York: W. W. Norton and Company, 1969.
22. U.S. Senate, Committee on Labor and Public Welfare. *Farmworkers in Rural America.* Ninety-second Congress, 2nd Session, 1972.
23. U.S. Commission of Civil Rights. *The Decline of Black Farming in America.* Washington, D.C., 1982.
24. Christian, V. L., et al. "Agriculture," in *Employment of Blacks in the South.* Ed. by R. Marshall and V. L. Christian, Austin, Texas: University of Texas Press, 1978.
25. Payne, W. C. *Implementing Federal Non-discrimination policies in the Department of Agriculture, 1974-1976,* Paper presented at the Agricultural Policy Symposium, Policy Studies Organization. Washington, D.C., 1977.
26. Coley, B. "Leaving the Farm," Southern Rural Development Center,*Rural Development Research and Education.* 4 (1977).
27. Doll, J. P., et al. *Economics of Agricultural Production, Markets and Policy.* Homewood, Ill.: Richard D. Irwin, Inc., 1968.

6

Small-Scale Agriculture and Rural Development

The economic and social health of rural areas in the U.S. is dependent upon the existence and economic well-being of small-scale farmers. Therefore, apart from the immediate effect of the individual farm families themselves, the constraints faced by the small-scale farms have also affected the development of rural areas in the country.

Small-Scale Farming and Rural Areas

The small-scale farm has been, and still is, an institution where the economic factors of firm and farm interface with the social factors of households and family in roles that coalesce to varied degrees of success in a bio-social environment. These factors consequently influence the rural community through important socializing activities in farm organizations such as Farmers Union, 4-H, REA, Farm Bureau and the Federal Land Bank.

Small-scale farms historically have created and stimulated the agricultural support base including the agribusiness complex that supplies production inputs to farming and provides market mechanisms for agricultural products. Some of these industries are located in rural communities providing off-farm employment and a tax base for the development of agriculture and the rural community.

For example, off-farm employment affects agricultural development through its impact on both the capital labor markets and through the services provided by rural communities to the farm population. Services that are provided to the farm population, especially primary and secondary education, are financed through local property taxes. In addition, off-farm employment sectors create additional savings which, to some degree, flow into local financial institutions. This undoubtedly strengthens the potential financial base of the rural community including the lending potential to the agricultural sector. The following statement by a rural bank official explains the importance of small-scale farms to the growth and development of rural communities.

> The rural community lives from the gross income of the family farm or the small, closely held family farm corporation. Because towns and banks are in the business of serving people, the banker sees that the disappearance of these families would cause his town and his bank to disappear...The fact remains that the small town can not exist without people on the land, no matter how productive a vast corporation farm may be (1).

Various studies have shown that the greater proportion of purchases made in local communities is by operators of small-scale farmers. For example, Marousek (2) reported that small farm operators in Idaho had a higher propensity than large farm operators to purchase both farm inputs and consumption goods locally. Most large farm operators buy their inputs in bulk from metropolitan areas or from large commodity groups.

Similarly, studies of the towns of Arvin and Dinuba in California, by Goldschmidt (3) and the Small Farm Viability Project (4), indicated that the community surrounded by small farms (Dinuba) had experienced a higher level of retail trade and greater growth rate of both retail trade and population than the community surrounded by large farms (Arvin). The small farm community also had about 2.5 times the number of independent business outlets found in the large farm communities.

A 1967 study by the Farmers Home Administration found that 190,000 farm families who had received credit from the agency grossed $3.2 billion, all of which was spent locally: $736 million for clothing, food and other consumer items; $1.7 billion for goods and services to produce corps and livestock; and $704 million to retire debt and buy new machinery (1). Therefore, a declining and distressed rural farm economy, due to constraints imposed on small-scale farms, means a declining economic base to support local retail and service establishments. In fact, local businesses today are vanishing at a faster rate than ever before, and it is estimated that, on the average, one small rural retail establishment is forced out of business every time six small-scale family farms leave the rural area (1).

At the same time that rural areas are enduring these hardships, federal support for rural economic development has been sharply reduced and many programs are facing elimination. A recent Ford Foundation report (5) stated that budget cuts of the Reagan administration have cost state and local governments in the rural South, where most of the small-scale farmers are found, a cumulative total of $20 billion since 1980. Many of these funds were critical to rural development efforts--providing jobs, technical support, infrastructural development, and business start-up capital for rural communities. Under the Gramm-Rudman-Hallings legislation, such programs will be cut another 8.5 percent this year.

In the past when the farm sector in the rural areas suffered a decline, communities survived due to the availability of manufacturing jobs. Now both are on a decline, particularly in the South, producing a crippling economy with large numbers of displaced and untrained workers. Even when some businesses are moving into the rural areas, they are more capital intensive and require fewer but well-educated and highly skilled workers. The rural area is without quality education or a skilled labor force and has a declining infrastructure (6). Given these facts, the rural areas will probably not attract any of the major sophisticated high tech and growth industries that some industrial policy proponents have identified as factors for the future revival of rural economies.

Rural economic activities are also greatly influenced by international markets and general economic policy issues such as interest rates and the value of the dollar (7,8). In recent years, international competition has worked against the rural areas. The fact that the U.S. competes in a truly world market has been rudely brought home to the rural people at a time when the country is losing the world markets, especially for farm products and manufacturing goods. When cheap labor is in demand, the U.S. cannot compete with developing countries, particularly those in Asia.

Because of the various combination of problems, many rural communities are today mired in economic stagnation and the trends for the future, unless something is done, are almost uniformly negative. As was the case in the past, the future economic problems of the rural areas will affect blacks more than it will whites.

Implications for Rural Areas

While the precise nature of rural development problems in the U.S. cannot be completely foreseen, the impacts of the constraints discussed in Chapter 5 are severe on those communities where agriculture continues to be the principle source of employment and economic activity. Small-scale agriculture directly contributes to jobs, wages (income), and the tax base for the rural community of the nation. Therefore, rural areas experiencing a distressed farm economy are faced with problems of unemployment, reduced income, and taxes to provide needed services. For example, rural governments and school districts are heavily dependent on property and sales taxes to provide basic services. A tax constraint due to a declining agricultural economy means a reduction in the ability of local governments and school districts to obtain revenues, hence, cash flow problems.

To solve the cash flow problems, these government units either have to cut expenditure and/or raise taxes, borrow funds at higher interest rates, or face bankruptcy. Although instances of government bankruptcies are quite rare, rural governments in areas where farm income and population is rapidly declining may go bankrupt because of insufficient revenues. Although for different reasons, in 1983 the San Jose, California, Unified School District declared bankruptcy, partly because of its inability to raise sufficient revenues following the enactment of Proposition 13, which limited property tax rates in the state. Furthermore, tax constraint will reduce the local government's ability to finance new public infrastructure to attract new firms and industries to locate in rural areas.

Besides small-scale agriculture's direct contribution to jobs, wages and the tax base for the rural communities, it has large "multiplier" or "ripple" effects on the agribusiness sector of the rural, state, and national economy and on those small businesses dependent on the small-scale farm economy. Farming provides jobs in agricultural input industries, agricultural processing industries, food and fiber wholesaling, and retailing industries. Some of these industries are located near farms or in the rural community. Thus, small-scale agriculture includes not only employment in farming, but in all businesses required to support the eventual delivery of food and other products to domestic and foreign consumers.

In addition, tax constraint or reduction reduces the fiscal flexibility of local governments and may force local governments to cut basic essential services. This will make the rural area less attractive to current residents and potential newcomers or migrants from cities where more essential public services are provided. In short, since rural areas are heavily dependent on property taxes, they are more vulnerable to the fiscal strain created by declining small-scale farm incomes and a distressed agricultural economy than other areas of the country.

A declining farm revenue will also mean a declining purchasing power base to support local retail and service establishments (laundries, dry cleaners, automobile dealers, banks, drug stores, restaurants). These businesses today are vanishing at a faster rate than ever before. In fact, some rural areas in the country, except for an occasional festival to celebrate one event or another, show little signs of daily life. Not surprisingly, often it is the older residents who have been greatly affected and left behind to fend for themselves in an environment of fewer services.

For the people who are left behind, the effect is not only fewer services but also increased costs for these services. However, most of these rural residents, with their income either constant or decreasing, cannot contribute indefinitely to the rise in costs of these services. As a result, further cuts may be made in some services which will result in destroying the economic viability of small-scale agriculture with its many consequences in the rural areas.

A declining small-scale farm economy causes more of these farm land owners to consider selling, and the inability of other new or local small-scale farmers to buy farmland makes it possible for increased acquisition by large-scale farmers, developers, and/or speculators. This, in turn, means further increase in concentration of farms or a decrease in lands devoted to farming with their commodity price implications.

Large farms use fewer total inputs per unit of outputs, and so a trend toward larger units will lead to a general reduction in economic activity in rural communities. For example, Tweeten (9) estimated that if American agriculture were reorganized into farms with sales of $200,000 and up, agriculture-related economic activity in rural communities would decline to about 78 percent of 1981 levels. On the other hand, if agriculture were reorganized into farms with gross sales of $20,000-$40,000, economic activity in rural communities would rise 5 percent over 1981 levels.

Another similar study (10) found that conversion of agriculture to larger farms would cause the amount of income generated in rural communities to fall by about 17 percent compared to a typical farm alternative, while conversion to small-scale farms would lead to an increase of about 14 percent. The concentration in farm wealth, in the hands of large-scale farms, some claim, will not stop when all farms now called small-scale farms have ceased to exist. Instead, they argue that unless small-scale farm programs and agricultural policies are "somehow turned around", the demise of farms with gross sales of $40,000 to $100,000 ".....will not be far behind"

(11). One supporting factor of these forecasts is the severe financial difficulty faced by this group of farmers in the 1980s.

Furthermore, as in the past, present and future technological developments will create a situation where it takes fewer and fewer farmers to feed and clothe the nation's population. For example, the present biotechnological developments in the farm sector promises further decline in the number of farms due to the efficiency gains it is expected to provide.

Unless it is offset by increase in domestic demand and/or agricultural exports, the increase in efficiency expected will result in fewer resources devoted to farming and fewer farmers. Such a substantial change in farming not only brings losses of individual sources of livelihood, for those who are engaged in farming and of the rural area's entire trade and service sector, but of a structure that has been a mainstay, a way of life in the history of American agriculture.

Implications for Agricultural Policy

The various economic and non-economic factors discussed in Chapter 5 have been major reasons for the problems that are faced by small-scale agriculture and rural areas in the U.S. These problems, therefore, are legitimate agricultural policy concerns for the nation. It is a concern because of the economic stagnation and disappearance of individual small-scale farms resulting from concentration and of the long-term trend towards further concentration of farm economic activity into so few hands which could result in food prices being raised to significantly higher levels.

A recent study predicts more concentration of farm wealth and production coupled with fewer farms in the year 2000 (12). If agriculture at some point becomes so dominated by large-scale farms, the very few institutional structures that now exist for marketing, input procurement and financing, which help small-scale farms to exist and keep the farm sector competitive, will disappear.

The future structure of farming is, therefore, an important competitive economic issue, and policies must be designed to maintain a competitive structure in the agricultural sector of the country's economy. There are many and varied proposed public policies for structural changes in the farm economy. Policies addressing the small-scale farm problems range from doing nothing to reforming the international and national monetary policies (which will not be addressed here) (13,14).

A do-nothing policy is workable, if the goal of the federal government is to speed up the structural changes in the farm economy. However, apart from effects on the distressed farm families themselves, a do-nothing policy has effects on the nation's economy (decrease in income tax), and the ability of local and state governments and school districts to obtain tax (property) revenue. The importance of small-scale farms contributions to the country as a whole, and particularly to the rural areas, have already been discussed in the previous section.

In addition, full time small-scale farms contribute significantly to the cheap but reliable food supply system of the nation. About 18 percent of the consumers' income in this country is spent on food, compared to countries in Europe who spend more than 30 percent of their income on food. A do-nothing policy is a threat to this cheap and dependable food system. Therefore, for this group of farms to survive, innovative policies and programs at both the state and federal level are needed.

References

1. U.S. Senate, Committee on Labor and Public Welfare. *Farmworkers in Rural America.* Ninety-second Congress, 2nd Session, 1972.
2. Marousek, G. "Farm Size and Rural Communities: Some Economic Relationships," *Southern Journal of Agricultural Economics,* 2 (1979).
3. Goldschmidt, W. R. *Small Business and the Community: A Study in Central Valley of California on Effects of Scale of Farm Operations.* Report of the Special Committee to Study Problems of American Small Business, Seventy-Ninth Congress, 2nd Session, Washington, D.C.,1946.
4. Small Farm Viability Project. *The Family Farm in California.* A report by the Technology Task Force, Sacramento, California, 1977.
5. Ford Foundation. *Shadow in the Sunbelt: Developing the Rural South.* A report of the MDC panel on Rural Economic Development, Chapel Hill, North Carolina, 1986.
6. Chicoine, D. L. "Infrastructure and Agriculture: Interdependencies with a Focus on Local Roads in the North Central States," in *Interdependecies of Agriculture and Rural Communities in the Twenty-first Century: The North Central Region.* Conference proceedings, Ed. by Peter F. Korsching et al., The North Central Regional Center for Rural Development, 1986.
7. Tweeten, L. G. "New Policies to Take Advantage of Opportunities for Agricultural and Rural Development," in *Interdependecies of Agriculture and Rural Communities in the Twenty-first Century: The North Central Region.* Conference proceedings, Ed. by Peter F. Korsching et al., The North Central Regional Center for Rural Development, 1986.
8. Henry, M., et al. "A Changing Rural Economy," in *Rural America in Transition.* Ed. by Mark Drabenstott, et al., The Federal Reserve Bank of Kansas City, Research Division, 1988.
9. Tweeten, L. G. "Agriculture and Rural Development in the 1980s and Beyond," *Western Journal of Agricultural Economics,* 2 (1983a).

10. Sonka, S. T., et al. *American Farm-Size Structure in Relation to Income and Employment Opportunities of Farms, Rural Communities, and Other Sectors,* CARD Report, No. 48, Center for Agricultural and Rural Development, Iowa State University, Ames, Iowa, 1974.
11. Carter, H. O., et al. *Research and the Family Farms,* a paper presented for the Experiment Station Committee on Organization and Policy. Cornell University, Ithaca, New York, 1981.
12. Schertz, L. P. "Farming in the United States," in *Structure Issues of American Agriculture,* Agricultural Economics Report No. 438, USDA, Washington, D.C., 1979.
13. Bullock, B. J. "Farm Credit Situation: Implications for Agricultural Policy," in *Economic and Marketing Information for Missouri Agriculture.* Cooperative Extension Service, University of Missouri--Lincoln University, 1985.
14. Tweeten, L. G. "Impact of Federal Fiscal-Monetary Policy on Farm Structure," *Southern Journal of Agricultural Economics,* 1 (1983b).

7

Programs, Policies, Research and Extension Needs of Small-Scale Agriculture

The expressed intent of American agricultural programs and policies, since the days of Thomas Jefferson, has been to support family farms, particularly small-scale land-owning farms (1). As a result, small-scale farms historically constituted the majority of American agricultural enterprises. Rural communities were and still are dependent upon this class of farms for their economic viability and survival.

However, for the past half century, a great many small-scale farms in the country have been lost. And with their loss, the viability of rural communities has declined. If the goal is to preserve the small-scale farm system as a way of life in the structure of the nation's agriculture, this trend must be stopped or reversed. To do so, policies, programs, research, and extension services must be initiated and directed to meet the specific needs of these farms.

Because of the diverse problems of small-scale farms, a single national or state program, policy, research or extension initiative will not have an equal effect on all small-scale farms. An effort must be made to address the full range of small-scale farm needs, including non-farm as well as farm development programs. Treating small-scale farms as a special needs group would allow the government to pursue more realistic as well as more effective programs. In fact, many programs would be far more cost effective if directed towards the specific needs of small-scale farms.

Programs, policies, research and extension needs to alleviate problems of small-scale farms and rural areas should, logically, be based on the constraints of small-scale farms identified in Chapter 5. They include problem areas of marketing, government commodity programs and policies, farm credit, technology, and education. The following are explanations of and suggestions for small-scale farm needs in these and other areas.

Technology Programs

Demand for appropriate agricultural technology by small-scale operators exists today and is likely to increase in the future due to rapidly rising production costs. But much of the farm machinery and equipment now on the market is not applicable to small-scale farm operation. Therefore, to increase the income from small-scale agriculture and reduce the cost of production, there is a need to focus on technology programs that should stress production of machinery and equipment which is sized to the needs of this group of farms; easy to maintain; free of unnecessary gadgets; and applicable to the labor situation. For example, a two-cylinder tractor that is fired with a magneto, uses no batteries, and is cranked with a flywheel is the kind of equipment needed by many small-scale farmers today.

In a recent survey of small-scale producers in the Northeastern region (NER) of the country, farmers were asked to describe new technology that would enhance the farms in their county. Of the eight technological needs they identified, machinery and equipment were ranked as the number one need by a majority of the respondents (2). The respondents also suggested some specific machinery and equipment technology pertaining to walk-behind tractors, economical but mechanical livestock feeding systems, small-scale mechanical tree fruit harvesters, improved and fuel efficient maple tree sap evaporators, and mechanical methods to apply lime and fertilizer economically on hillside pastures.

Underlying most of the above suggestions was the need for technology focusing on smaller, more durable multipurpose machinery and equipment that would be economical to operate. The technology is available to develop such machinery and equipment and sell it at a reasonable price. An incentive, however, must be provided for agricultural equipment manufacturers to consider the production and sale of such technology. This could be achieved by giving manufacturers of farm equipment investment credits of a higher amount than present levels.

Escalating energy costs is a major concern of small-scale producers. Therefore, alternative technology should also focus on energy production by the small-scale farmers for self-sufficiency and reduced energy requirements. For example, systems utilizing a minimum of fossil fuels should be exploited. Most of the survey respondents above suggested energy production technology from methane gas and alcohol from waste products generated on the farm. They further suggested increased efforts to be made in utilizing forest products, wind, and sun as energy sources (2).

Small-scale technology is needed for small-scale producers to operate their farms more intensely and to maximize returns from every item grown or produced. It will be essential that on-the-farm energy sources be developed for small-scale producers and that they be practical for their farm business. In short, any new technological breakthroughs in machinery, equipment and energy efficiency for small-scale farms will not only lower production costs and increase their income, but enhance the well-being of the rural, state and national economy.

Management and Information Programs

In the future, small-scale farms can compete with larger commercial farms, only if they are better managed. Therefore, policies and programs directed toward these farms may be appropriately directed toward improving their managerial ability. More importantly, they need to be given every opportunity to learn to manage the farm as a business operation that it must become to survive. Greater emphasis, therefore, must be placed on research and extension programs to improve the management ability of these small-scale farmers. One way this could be achieved is for the institutions to be active in on-the-farm demonstration, or what is today known as Farming System Research/Extension, to show the economic value of various management practices.

Planning alternative and high value crops, more simplified record keeping procedures, financial and credit alternatives, pricing mechanisms and approaches, and feasibility of off-farm employment are some examples of the areas of management needs of small-scale producers. Additional management needs of the small-scale producers are easy access to records or computerized information available at educational and research centers such as the Land-Grant universities. Examples of such needs are computer programs that are scaled to small-scale farms to identify alternative enterprise selection determined by profitability of the enterprise.

Information distribution and more efficient communication systems were identified as of paramount importance by small-scale farmers of the NER of the country (2). Demonstrations focussing on small-scale agricultural enterprises; publications with more basic pictorial type; tours to successful small-scale farms; computerized budgets and program information on finance, production and marketing that are functional on the small-scale farms are examples of ways to meet these information needs.

Most available information in the past was not suited to the needs of small-scale producers. Therefore, in any new development of an information system, its usefulness by this group of producers must be considered. That is, there is a need for development of an information system that will reach out and help the small-scale producers so that they are better able to make sound farm business decisions. This is particularly true today more than in the past, because of the information age which is coming towards us with phenomenal speed and the wide array of new bio-technologies that are expected to revolutionize animal and plant production in U.S.

agriculture (3,4,5,6,7). In short, there is a need to develop organizational and technical management assistance that is tailored to meet the needs of small-scale farmers so that they have precise information when and where it is needed for making a sound farm management decision in marketing, production and the purchase of inputs.

Marketing Programs

Small-scale farmers do not produce enough output, hence, they are not in a position to benefit directly or indirectly from the marketing practices and pricing systems that exist today. Therefore, there is a need to develop alternative marketing systems for the small-scale farmers to market their products.

In recent years, the interest in marketing by small-scale producers has centered around the direct farm-to-consumer marketing alternatives. These markets are identified as road-side, farmers', pick-your-own and pooling or cooperative marketing operations. These marketing alternatives have particularly become increasingly viable in the South as industry moves into rural areas and population density increases.

For example, vegetable farmers may become more efficient by producing only one type of vegetable and selling it at a centralized (cooperative) marketing system. The efficiency of a centralized marketing system comes from limiting the transactions made by each farmer, and providing sufficient quantities of high quality produce through cooperative markets to get the attention of large-volume buyers. Possibly 10 to 20 small-scale producers marketing cooperatively could develop these kinds of new markets.

Furthermore, new markets could be developed in areas where major retail outlets are not concentrated. These markets could be neighborhood markets in areas of low-income, large population with minimum or no transportation availability. Also major business centers such as industrial plants, office buildings and construction sites may provide direct market opportunities for small-scale producers of fruits, vegetables and other produce.

The direct-marketing systems described above are the simplest means of selling products for many small-scale farmers, but they have their limitations. For example, advertising and promotion is required for success in some of the marketing which may add cost to the product. If a small-scale farmer growing fruits and vegetables has a pick-your-own operation, he will probably need to advertise. He may use road-side signs, ads in the newspaper, radio, television or just send postcards to all customers that will come to pick-your-own operations. These kinds of basic marketing processes have to be considered both in terms of cost and revenues. Another limitation is that direct-marketing takes time and effort to develop. Although the development of direct marketing takes time and effort, it has proven to be profitable for some small-scale farmers.

An additional marketing alternative that is available but needs further development is advanced marketing such as cash forward contracting, hedging in the futures market, and options trading. These market alternatives can provide the small-scale farmers opportunities to bargain for a better price, reduce risk, and establish floor prices (i.e., a contract can be made to obligate the buyer to pay the farmer no less than a specific price, but should there be a further price rise, then the buyer must pay the farmer the higher price).

Furthermore, state and local governments could create more aggressive and innovative programs and policies to help match small-scale farmers with both buyers and consumers. Increasing funds for state and locally sponsored marketing efforts, training and improved information systems, technical assistance to producers groups, and improved sales strategies are ways to accomplish these objectives. In fact, such state and locally sponsored marketing programs will serve, and serve well, this group of farmers who have not traditionally benefited from the national marketing system.

State and locally sponsored crop diversification programs is another way to help improve the marketing problems of small-scale farmers. Such a program will provide encouragement and incentives for the small-scale farmer to grow products which will meet the demands of new, emerging markets. That is, rather than producing traditional crops, such as wheat for export or tobacco, a small-scale farmer today should be provided with incentives to grow vegetables, or blueberries and strawberries.

Restructuring of markets to reduce the current concentrated marketing system is an additional marketing program that would meet the needs of small-scale farmers. Such a restructuring of the market not only provides market access to small-scale producers but also creates a more competitive market that is beneficial to the consumers. Although some alternative markets are available to day for use by the small-scale producers, as clearly identified above, there is a need for more work to develop or find market outlets for these farmers to market and multiply their volume of sales, hence their income.

Education and Training Programs

Education and training skills have a direct impact on the farm manager's ability to make a sound management decision about adoption of alternative technology, marketing, and information systems. However, small-scale farmers lack the quality and quantity of education and training to make profitable decisions in their farm business. There is, therefore, the need to improve the quality and quantity of education and training received by these farmers.

For example, using the Farming Systems Research/Extension approach, there is a need for pilot or working small-scale farms to teach small-scale producers to consider innovative and appropriate technology for small-scale agriculture. Such an educational system would provide on-the-site training via involvement and observation of area small-scale farm operators. Furthermore, there is a need for a collaborative educational approach which involves farmers in multi-interest committees for problem identification and finding solutions.

There are also educational training needs in the areas of farm business management, record keeping procedures and use of computers so that small-scale farmers can be more efficient and effective farm managers. In fact, education will continue to be more essential and critical for meeting the new technology choices that small-scale farmers will face in the future.

Government Programs and Policies

U.S. agricultural policies and programs for the most part have not specifically benefited small-scale farmers, since they were mainly directed towards the needs of large farmers. If the goal is to preserve the small-scale farmers, programs and policies or assistance from the government must be better targeted to these farmers who are in real need of aid. Targeting of programs or policies in themselves is not enough; rather they should be designed in such a way as to maximize the number of people in small-scale farming able to sustain a living in the agricultural industry. This is particularly true in light of the average age of small-scale farm operators and their level of marketable skills (education) which will force many of these farmers to remain on their farms even if their relative income stays at its present level.

There is a need for programs and policies that are needs-based rather than production-based, as was the case in the past. An income support program is one example of a program that would vastly improve the distribution of benefits and would measurably improve the distribution of net income among small-scale operators. According to a 1986 research report (8), the income support aid to this group of farmers would provide a more equitable distribution through the application of a formula based on farm product sales. Under the formula, subject to certain upper and lower limitations of farm sales, each operator would receive a supplemental income support payment from the government to ensure that he/she has an amount of non-farm income and government aid equal to the average total of such resources available to his/her domestic competitors (i.e., all other farm operators in the U.S.)

The bio and information technologies discussed above will keep pressure on agriculture to expand products and product markets, particularly export markets. Farm policies and programs that will be developed to address these technological changes, should be sufficiently flexible and balanced to warrant the participation and share of potential benefits from these technologies by small-scale farmers. Otherwise, as was the case in the past, there will be a continuation and potential acceleration of the demise of small-scale agriculture if present and future farm policies and programs continue to ignore the needs of small-scale farmers.

Farm Credit Programs

For the various reasons discussed in Chapter 5, small-scale producers have been, and still are, plagued by credit problems. However, without an adequate source of credit, they can not invest in new technologies or expand their farm operation to increase production and improve their farm income. One of the major reasons for this situation is the conservative lending practices of credit institutions, which prefer helping large-scale rather than small-scale farms. Therefore, there is a need for restructuring the conservative lending institutions' policies which prefer to lend to profitable farmers. One way to help speed up the restructuring may be to provide tax relief for the credit institutions if they increase their loan activity to small-scale producers.

Small-scale producers need access to capital markets at terms and interest rates that are reasonable and affordable. State or federal government providing direct low-interest loan programs is one solution to this problem. And these credit programs must be specifically designed to help small-scale producers finance needed changes in their resource endowments. In fact, credit programs combined with education and management technical assistance could be very successful in increasing productivity and the income situation of small-scale producers. This in turn would contribute to the growth and development of rural America.

The Farmers Home Administration was established to service the credit needs of farmers who failed to meet the lending criteria of other institutions, but the program has failed to advance, and in some cases, might have hindered the efforts of small-scale farmers to remain in farming business. Therefore, there is a need to redesign this institution's farm credit program so that it can help the small-scale farmers meet their finance needs. In any new credit program, a great deal of attention must be given to the unique capital and cash flow-limiting factors normally associated with this group of farmers who are often not in a position to take advantage of other farm programs such as price and income supports (9).

Effective Research and Extension Programs

Research and extension activities basically should be for all farmers. However, large-scale producers, with the necessary skills and finances available to them, have largely benefited from technologies developed by private and public research agencies more than have the small-scale farmers. In fact, if favorable prices for inputs and outputs remain, private agricultural input and marketing research agencies will continue to develop and disseminate technologies that will further enhance the production and marketing of the large-scale producers. Therefore, the main effort of current and future research programs, particularly of public agencies, must be directed towards addressing the problems of small-scale producers and finding solutions to their problems.

Research Programs

For the past few years, the priorities of some public research agencies in the nation have been to address the problems of small-scale farmers and their impact on rural areas. However, for the most part the focus of the research efforts has been predominantly one-dimensional and designed to solve a single problem of the small-scale producer. A research effort to increase crop yields without considering the cash-flow, labor requirements and other factors of the farm is an example of such research. However, the problem of small-scale farmers is multi-dimensional and the on-going research is inadequate to deal with the many interrelated aspects of the current small-scale agricultural problems.

Therefore, there is a need for a multi-dimensional, interdisciplinary and interagency research effort to solve the problems of small-scale producers in the country. A comprehensive systems approach to research efforts of production, management, marketing, policy, credit and off-farm employment opportunities is such an example. In addition, any research that addresses the small-scale farm problems must incorporate the development of the rural areas. This is particularly essential since most of this group of farmers reside in rural communities which are heavily dependent on revenues from small-scale agriculture as well as serving as the off-farm job market for the small-scale producers.

The survey of small-scale producers of the NER clearly indicates an immediate need for the integrated systems approach to research as a solution to their problems (2). Table 7.1 represents suggestions from these producers for some areas of specific and immediate research that might help them. The majority of the respondents suggested a need for research to assist their production capabilities, followed by a need for management and marketing research. Suggestions for immediate research into production, management, and marketing totaled 68 percent, indicating the importance of identifying fundamental problems of the farming business for small-scale farmers.

Furthermore, research has to be geared toward developing crop alternatives that will blend well with off-farm job opportunities, as well as provide a comparative advantage to the small-scale farmer. Research specifically directed to vegetable and fruit crops benefit this group of farmers. Specific research to develop these crops to be more tolerant to factors such as exposure to light frost, cold weather, drought, and pollution are some examples. Also, there is a need to control weeds, insects and the diseases of vegetables and fruit crops.

TABLE 7.1: Suggestions for Specific and Immediate Research by Small-Scale Farmers of the NER

Areas of Research	Suggestions	
	Number	Percent
Production	37	27.8
Management	30	22.6
Marketing	24	18.0
Economics	16	12.0
Disease and pest control	9	6.8
Equipment	6	4.5
Quality and preservation	2	1.5
Energy	2	1.5
Other	7	5.3
Total	133	100

Source: *A Survey of Current and Expected Research Needs of Small Farms in the Northeast Region,* Science and Education Administration, USDA, 1980.

In addition, research into how to grow the best and highest yielding transplants in the shortest time; production systems for up to four crops per year on the same parcels of land; and cultural practices to reduce labor requirements for the small-scale producer is needed. Since vegetables and fruit crops are sold fresh, research is needed to develop varieties with improved flavor and eye appeal which would further benefit the small-scale producer.

Extension Programs

The success of new research discoveries being quickly absorbed and implemented to benefit the small-scale producers is dependent upon extension program priorities for this group of farmers. In the past, the extension service has devoted much effort toward educating farmers on the advantages of adopting the latest recommended production, cultural and managerial practices. These efforts have resulted in increased yields, improved quality and increased income primarily for the large-scale producers. The service has achieved little success with the small-scale producers. Therefore, there is a need to increase the efforts of the extension service to help the small-scale producers to increase the profitability of their agricultural enterprises.

Past research has found that small-scale producers had less contact with extension agents due to the reluctance of this group of farmers to solicit assistance from the agents and, in any event, the first contact priorities by agents were mostly given to large-scale producers (10,11,12,13). In recent years, the extension service has recognized the needs of the small-scale agriculture clientele, but has failed to implement a more aggressive program to help small-scale producers adopt current agricultural practices.

According to the NER small-scale producers (2), for the extension service to ultimately be more effective in fulfilling the needs of this group of farmers, its priorities should be directed towards contacts, aggressive outreach programs and improved information systems (Table 7.2). As was the case in the past,

TABLE 7.2: Suggestions for Effective Extension Programs by Small-Scale Farmers of the NER

Extension Program	Suggestions Number	Suggestions Percent
Contacts or aggressive outreach	27	25.5
Improved information systems	19	17.9
Economic studies and guidelines	13	12.3
Education and training	12	11.3
More involvement of small farm operator	12	11.3
Agrifarm organizations established	8	7.5
Identify and rectify the needs	8	7.5
Other	7	6.6
Total	106	100[a]

[a]Due to rounding, total may not add up to 100.

Source: Ibid.

it is hard to reach the small-scale producer by conventional approaches such as field days, group meetings and mass media methods. As indicated by the majority of the NER small-scale producers in Table 7.2, there is a need for a more aggressive outreach effort and the development of new information systems by the extension program to help small-scale producers.

These needs are particularly greater today than in the past due to the increasing number of small-scale farmers (and most are in the older age brackets) with full time off-farm jobs. As the number of off-farm jobs increases the extension service will have a problem contacting and working with these farmers during regular working hours. Therefore, it must design new information systems to combat this problem and any strategy it develops should take into consideration the differences in the age and education levels of this group of farmers.

Special education and training programs by the extension service are also needed to help the small-scale producers. The educational programs need to be designed utilizing the Farming Systems approach. The traditional disciplinary or component approach to extension education has been largely inadequate in dealing with the many interrelated aspects of the current small-scale agricultural problems. Through an integrated multi-disciplinary systems approach, extension educational programs will likely address more effectively the needs of small-scale producers.

For example, the development and implementation of a comprehensive farm plan using available resources of the small-scale producers would be more successful if there is a combined effort from all disciplines of the service personnel. In addition, there is a need for special education and training programs for some of the extension agents who apparently find it difficult to deal and cope with small-scale producers' problems.

As indicated in Table 7.2, more involvement with the small-scale farm operators is another area of priority that needs to be addressed by the extension program. Such a partnership between the extension service and farmers will have several implications for extension programs. First, extension programs for small-scale producers are more effective in bringing about changes when the clientele are involved in a collaborative relationship. Second, the generation of appropriate or relevant technology and/or identification of alternative agricultural enterprises is more expedient when extension agents work with the farmers as an integrated multi-disciplinary team. Third, the extension agents will become more aware of these farmers' needs and relate to these needs more effectively. Fourth, it will be easy for the agents to gather data directly from the farmers when developing extension programs. Fifth, the farmer collaborators complement the efforts of the extension agents for the diffusion of physical, social and economic changes that are taking place in their community (2).

Rural Development Programs

In the past, occasionally public concern has risen over the plight of small-scale farms and the operators and families associated with these farms. A public policy arena, however, should go beyond the area of agricultural production and marketing to include topics of community development and social service delivery. To focus on programs specifically for agriculture and ignore the non-farm sector would not be in the best interest of small-scale farms.

Obviously, many small-scale farmers will continue to be dependent on off-farm income for an adequate level of living and survival. Experience has shown that some small-scale farmers have successfully combined farming with off-farm jobs to improve family income and better utilize family labor. However, many lack the skills needed for the kind of off-farm jobs available; they also lack information about job opportunities.

Therefore, small-scale farm programs today should continue to focus on the issues of farm family poverty and alternative ways to improve the well-being of the poor farmer. Non-farm options for retaining and maintaining and improving the well being of small-scale farm families should be included in future small-scale farm programs. Ideally, industries or other sources of non-farm employment, which provide off-farm employment opportunities that would mesh well with the objectives and survival strategies of small-scale farmers, including part-time farming and full-time off-farm jobs, should be brought in. In fact, those farmers who combine farming with non-farm activities appear to have staying power. The greatest loss in the number of farmers has occurred among those who did little off-farm work. Thus, any predominantly farm-option-only approach should be seriously reconsidered because the farm is no longer the sole source of income for many farm families.

It is evident that an integrated rural development program for the entire rural community focussing attention on the provision of social services and non-farm job opportunities is needed. Rural enterprises that can create additional income and off-farm job producing potential can be initiated. A model organization that can identify new product ideas, provision of non-traditional specialty commodities, or seek new small business ventures with potential that fit the depressed rural community environment and are capable of being structured and nurtured with technical and financial support from the federal and/or state governments can be developed to improve the well-being of rural residents.

The study by the Office of Technology Assessment (9) suggests some specific policy options to ease the strain on rural communities caused by structural changes in agriculture:

1. initiation of comprehensive programs for community redevelopment and change throughout rural America.
2. revitalization of rural community development research and extension programs to serve the needs of communities in transition.
3. increase employment opportunities in rural areas by aggressively attracting new business activities to rural communities, with particular emphasis being placed on attracting those business that develop technologies and serve the need for high-technology agriculture in rural areas.
4. assisting rural communities in developing and modernizing the infrastructure needed to be a socially and economically attractive place to live.

5. encouraging and/or providing for rural communities to play a vital role in skills training for displaced farmers and rural community employees.
6. modifying school, college, and university outreach programs to serve the important role played by rural communities.

It will take a lot of time and a considerable amount of effort by federal and state governments, and rural people, to revitalize small-scale agriculture and rural communities, but it can be done. As a matter of fact, it is being done in some communities across the country. An example of this can be seen in some small towns in North Central Missouri, which had lost large numbers of manufacturing jobs, were not attractive to new industry, had large numbers of elderly, a faltering farm economy and losses of both people and income to other areas (14). But these small towns today are surviving because of what the townspeople accomplished by working together to develop means that worked to create jobs and other economic activities. Furthermore, a recent publication from a national extension workshop that was held in Minneapolis, Minnesota, documented how small rural communities in some parts of the country are utilizing tourism and travel as a rural development tool and provided new ideas on how other rural communities can expand businesses and services for tourists and travelers (15).

In summary, small-scale farm families and rural people, in general, are interested in assistance in making small-scale farm and rural areas more viable, with a higher quality of life. Furthermore, the family and the small-scale farm institution are viewed by many to be important to society's strength. Therefore, the preservation of small-scale farms should be the central feature of American agricultural policy, because it is a society and a way of life upon which rural America is based.

Special Programs for Black Farm Operators

The economic condition and public policies that affect the small-scale producer also have an impact on the viability of black and other minority farm operators. However, the impact on black farm operators is greater than the impact on small-scale producers in general due to poverty, lack of education and institutional barriers for black farmers.

The agricultural resources of these farmers are inadequate and/or managed poorly, and, as a group, they also face problems in securing off-farm jobs either because of the lack of education and training, or age or race. As a result most black farmers receive negative net farm income and total family income below that of the non-metropolitan median household income. Therefore, there is a need to develop special programs and policies to address the problems of black farm operators.

Policy concerns of black farmers involve economic security and equity issues. For years, most black farmers were not allowed access to the institutions designed to service farmers, and like other rural blacks were unable to capture their proportionate share of the general rural growth process. Strategies to improve income proved difficult to implement when institutions were unresponsive to the circumstances faced by blacks.

To be successful, policies and programs implemented to improve income security have to be supported by efforts to assure and improve access to responsible institutions. Thus, for the remaining and declining number of black-operated farms, government policies stemming from the 1964 Civil Rights Act and Equal Employment Opportunity programs should have an important role in assisting black farm operators by directly affecting the social climate under which assistance programs operate.

Special programs are also needed for the black farmers and black rural communities in all areas of social and economic needs and services. Special programs to assist blacks are needed in marketing the use of chemicals, resource management, record keeping, cash flow planning, law and regulations affecting agriculture, land ownership and tenancy situations, patterns of tillage, economic organizations, participation in local decision making and policy formation, management of food related resources (such as nutrition, practice in food storage, safety and sanitation), and any public and social services that would promote community development and retention of black farmers in the rural areas.

In addition, there is a need by both extension and research to play a greater role in training and improving the managerial capabilities of black farmers. A special program may be needed initially to establish contact with black operators and involve them in special educational activities. Another special program need for black farm operators is the development of better methods of educating and disseminating practical and useful information. Particularly, special attention must be given to their educational needs to stop or at least to slow down the rate of decline, as well as improve the well-being of this group of farm operators.

References

1. Maddon, J. P., et al. "Toward an Agenda for Small Farm Research," *American Journal of Agricultural Economics,* 61 (1979).
2. U.S. Department of Agriculture. *A Survey of Current and Expected Research Needs of Small Farms in the Northeastern Region.* Science and Education Administration, Agricultural Research Results, 1980.
3. Masuda, Y. *The Information Society.* Tokyo, Japan: Institute for the Information Society, 1980.
4. Dizard, W. P. *The Coming Information Age: An Overview of Technology, Economics and Politics.* New York: Longman, 1982.
5. Dillman, D. A. "The Social Impacts of Information Technologies in Rural North America," *Rural Sociology,* 50 (1985).
6. Knutson, R. D., et al. "Technology as a Force of Change," in *The Rural Great Plains of the Future.* Symposium Proceedings, Denver, Colorado, 1987.
7. Molnar, J. J. "Biotechnology and the Small Farm: Implications of an Emerging Trend," in *Strategy for Survival of Small Farmers: International Implications.* Ed. by T. T. Williams, Human Resources Development Center, Tuskegee Institute, Tuskegee, Alabama. 1985.
8. U.S. Congress, Subcommittee on Agriculture and Transportation. *Selling Out the Family Farm: A Classic Case of Good Intentions Gone Awry.* Ninety-Ninth Congress, 2nd Session, 1986.
9. U.S. Congress, Office of Technology Assessment. Technology, Public Policy and the Changing Structure of American Agriculture. OTA-F-285, Washington, D.C., 1986.
10. Huffman, W. "Black-White Human Capital Differences: Impact on Agricultural Productivity in the U.S. South," *American Economic Review,* 71 (1981).
11. Vogeler, I. *The Myth of the Family Farm: Agribusiness Dominance of U.S. Agriculture.* Boulder, Colorado: Westview Press, 1981.

12. Baldwin, S. *Poverty and Politics: The Rise and Decline of the Farm Security Administration.* Chapel Hill: University of North Carolina Press, 1968.
13. Janvry, D. A., et al. "Toward a Rural Development Program for the United States: A Proposal," in *Agriculture and Beyond: Rural Economic Development.* Ed. by Gene F. Summers and et. al., University Wisconsin, Madison, 1987.
14. McCally, J. *Small Town: Survival Manual.* Manual No. 133, Northwest Missouri, University Extension, University of Missouri, Columbia, 1988.
15. *Using Tourism and Travel as a Community and Rural Revitalization Strategy.* Proceedings of the National Extension Workshop, Ed. by John Sem, et al., Minnesota Extension Service, University of Minnesota, St. Paul, Minnesota, 1989.

Index